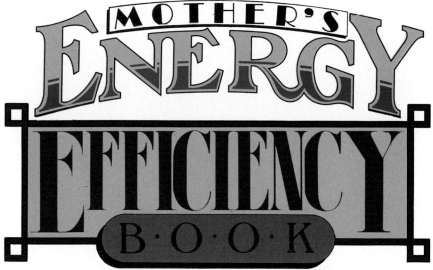

HEAT, LIGHT, POWER

by the editors and staff of
THE MOTHER EARTH NEWS
Project Editor: William C. Davis
Art Director: Wendy Simons

THE MOTHER EARTH NEWS, Inc.
Hendersonville, North Carolina
Distributed by Brick House Publishing Co., Inc.

Published 1983
Printed in the United States of America
THE MOTHER EARTH NEWS, Inc.
105 Stoney Mountain Road,
Hendersonville, North Carolina 28791

Library of Congress Catalog Card Number 83-62283
ISBN 0-938-43205-2

THE MOTHER EARTH NEWS®

magazine is published at 105 Stoney Mountain Road, Hendersonville, North Carolina 28791. It is the bi-monthly publication edited by, and for, today's turned-on people of all ages . . . the creative ones, the doers, the folks who make it all happen. All material in MOTHER'S ENERGY EFFICIENCY BOOK: HEAT, LIGHT, POWER has previously appeared in, or has been adapted from, THE MOTHER EARTH NEWS magazine or a related publication. Not all facts and figures are necessarily current.

This book was produced by THE MOTHER EARTH NEWS® magazine staff, with special acknowledgment to the following: Bruce Woods, Arthur Snell, Barbara S. Henderson, Lorna K. Loveless, Donald R. Osby, Richard Freudenberger, David Schoonmaker, Jennifer Fisher, Michael Garner, Anna Gravley, Kathy Ochsner, Charles R. Milstead, Marsha Drake, Steven Keull, Ken Cross, Ken Forsgren, Jack Green, Carolyn Dellinger Sizemore, Sherry Seagle, Carolyn Frederick, Spanky Alexander, Jerry McMillan, Kathleen Seabe, Mark Hillyer, and Garry W. Ramo.

TABLE OF CONTENTS

INTRODUCTION

Energy surrounds us. It's in the air, in the water, in the sunlight, in the crops we grow, and even in our wastes. And harnessing it is easier and less expensive than most people imagine. Being unaccustomed to utilizing these resources, a great number of families continue to depend on expensive energy purchased from outside sources and piped into their homes.

Since 1973, however, when the prices of oil, natural gas, and coal took the first of several gigantic leaps upward, many folks have been experimenting with new methods and reviving old ways of manufacturing their own power from the environment around them. Quite a few of these inventive people have been working within limited budgets, which mandated that their designs be inexpensively and easily built.

Some folks turned to the sun, installing solar collectors to capture the heat for warming their homes or heating their water. Often these units are passive, operating solely on the principle that hot air or water rises, creating convection currents.

Other people strove to reduce their need for visiting the gas pumps to fuel their cars. They investigated the possibilities of using alternatives to high-priced petroleum fuels, and then began manufacturing alcohol and making the minor adaptations needed to allow the homemade brew to burn in their standard internal combustion engines.

Farmsteaders and homesteaders with streams on their land began to put the flowing power source to work producing electricity for home use, pumping water for drinking or irrigation, and performing various chores to which rotational energy is applicable.

The wind was also put to work, and wind generators and pumps saw a resurgence of the popularity they enjoyed in the 1930's, '40's, and early '50's . . . before electric lines reached the sparsely inhabited parts of the country. Experiments involving nontraditional designs that operated on a vertical axis led to the use of alternatives to large towers and propeller-driven machines.

Interest in the areas of waste disposal, fuel production, and organic gardening methods helped to initiate a system for producing methane-rich biogas, which could serve the same functions as natural gas or could be used to fuel engines that ran electrical generators. Human, animal, and vegetable wastes were fed into a digester that produced biogas, leaving a nutrient-rich sludge that made an excellent fertilizer.

Cutting heating costs became important, and the woodstove made a new appearance, showing up in homes all across the country. The woodburners' popularity spurred research into improving their efficiency and safety, while reducing their production of polluting emissions. Alternatives to the conventional cast-iron devices were also tried, some incorporating large thermal masses that could store heat and then give off warmth over a period of several hours.

Individuals who wished to retain conventional central heating and air conditioning in their homes found the heat pump would reduce their heating and cooling costs. This led to improvements in these devices that made them more versatile and efficient.

And that was not the end. People found they could build instruments to raise and lower their thermostats automatically, that they could use their fireplaces to help cool their homes, that they could devise ways of obtaining additional heat from their stovepipes, and more.

Efficient conservation and utilization of the energy in the air, in the water, in the sunlight, and in our wastes involve choosing the best methods and equipment for a given situation. The energy itself is there. We are surrounded by it.

STAR POWER

The primary source of all the earth's available energy is the central star of our solar system, the sun, which provides heat and light for all its orbiting planets. Without the solar forces that struck the face of the earth millions of years ago, there would be no coal, oil, or natural gas at present, but those resources that took eons for the sunlight to help create are now in limited supply and are expensive to obtain.

However, the sun's rays that fall upon the earth today, and those that will reach our planet tomorrow, are plentiful and virtually free. This solar energy can be harnessed by individuals and used to warm their homes and water, to preserve their food by drying it, and to cook their meals. It can even be changed directly into electricity.

Better still, each of these tasks can be accomplished economically and easily, and without technologically complicated equipment and devices. In most instances, in addition to a little imagination and ingenuity, a homemade solar collector designed for the application in mind can enable a person to take advantage of this radiant energy and thereby to reduce his or her dependence on commercial sources of power.

STAR POWER

The Greedy Heat Grabber—
A Solar Room Heater That
Can't Help But Work

An inexpensive solar panel can keep a whole room warm and cozy on a sunny day, no matter how cold it is outdoors.

SOLAR COLLECTORS & WATER HEATERS

Do you shudder in dread when the climatologists predict that the next winter may be even colder than usual? Do forecasts of "clear but subzero" make you want to huddle under the quilts all through January and February? Well, your *indoor* temperatures need not reflect that chill because there's a fairly easy way, you see, to bring more warmth from the sunshine right into your home (without costly fuel bills) . . . *if* your house or apartment has one or more unshaded south-facing windows and *if* you outfit those windows with a homemade heat grabber.

A heat grabber is no more than a weathertight box that's insulated on the bottom, sides, and back, and faced with glass. An insulated divider is positioned inside this box and brought out its top to form an open "lip" at the box's upper end. This lip is designed to hook over a windowsill so that the window itself can be pulled down snugly onto the glass which covers the top of the heat grabber, leaving the main body of the solar collector "leaning against" the south side of the house at a 45°-or-better angle.

The operation of the unit is just as simple. When the sun shines, its rays pass through the glass on top of the heat grabber, strike the upper surface of the divider (which is painted black), and warm the aluminum foil covering on that divider. As the foil heats up, it in turn—warms the air next to it. And that air, as might be expected, rises up the face of the divider and begins to pour out the opening at the unit's top.

But, of course, that hot air can't move up the face of the divider unless it pulls cool air around the divider's foot to take its place. This pulls even more cool air in through the *lower* opening at the collector's top (the only place that cool air can enter the otherwise airtight unit) and down under the central divider.

A heat grabber, then, is a "convective loop" solar room heater that operates automatically on nothing but the sun's energy. Whenever the sun shines, this clever little unit just sits there happily pumping thousands of Btu into the house. And when the sun quits shining? The air in the box cools and tries to sink to the collector's foot . . . which "shuts off" the whole convective loop. (The heat grabber, in other words, will spew heat into the room when the sun shines . . . but it won't pull heat from the room when the sun doesn't shine.)

This simple and effective solar collector can be fabricated in just under an hour by an experienced home craftsman (or in less than two hours by the more fumble-fingered among us) for an astonishingly low price. And once constructed, this sturdy unit should give years of dependable service.

The key to the heat grabber's quick assembly and low cost is a rigid polyisocyanurate foam insulation board. This board is impregnated with glass fibers for strength, faced on both sides with heavy aluminum foil, and available in thicknesses ranging from 3/8" to 1-7/8". The material is marketed as a replacement for the pressed fiber sheathing or "blackboard" now used by contractors in the construction of wood-framed houses, and it's not recommended for any other purpose. Heat and other tests run on the insulation board, however, indicate that it's nearly ideal for use in quick, easy, and low-cost solar collectors such as the heat grabber. (Woven fiberglass duct board that you've wrapped in aluminum foil can be substituted.)

These two easy-to-make knives will cut through rigid polyisocyanurate insulation board without piercing the foil covering.

STAR POWER

The Greedy Heat Grabber—
A Solar Room Heater That
Can't Help But Work (continued)

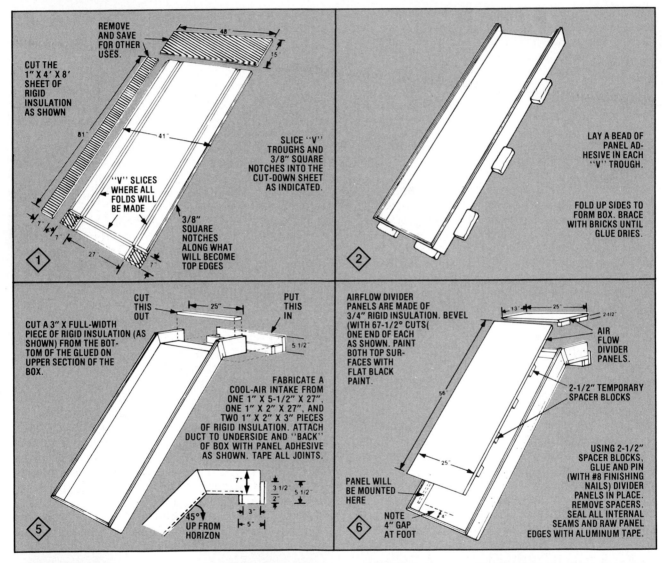

Panel 1:

CUT THE 1" X 4' X 8' SHEET OF RIGID INSULATION AS SHOWN

REMOVE AND SAVE FOR OTHER USES.

48"

15"

81"

41"

"V" SLICES WHERE ALL FOLDS WILL BE MADE

7"

7"

27"

7"

3/8" SQUARE NOTCHES ALONG WHAT WILL BECOME TOP EDGES

SLICE "V" TROUGHS AND 3/8" SQUARE NOTCHES INTO THE CUT-DOWN SHEET AS INDICATED.

Panel 2:

LAY A BEAD OF PANEL ADHESIVE IN EACH "V" TROUGH.

FOLD UP SIDES TO FORM BOX. BRACE WITH BRICKS UNTIL GLUE DRIES.

Panel 5:

CUT A 3" X FULL-WIDTH PIECE OF RIGID INSULATION (AS SHOWN) FROM THE BOTTOM OF THE GLUED ON UPPER SECTION OF THE BOX.

CUT THIS OUT

25"

PUT THIS IN

5 1/2"

FABRICATE A COOL-AIR INTAKE FROM ONE 1" X 5-1/2" X 27", ONE 1" X 2" X 27", AND TWO 1" X 2" X 3" PIECES OF RIGID INSULATION. ATTACH DUCT TO UNDERSIDE AND "BACK" OF BOX WITH PANEL ADHESIVE AS SHOWN. TAPE ALL JOINTS.

7"

3 1/2"

2"

5 1/2"

45° UP FROM HORIZON

3"

5"

Panel 6:

AIRFLOW DIVIDER PANELS ARE MADE OF 3/4" RIGID INSULATION. BEVEL (WITH 67-1/2° CUTS(ONE END OF EACH AS SHOWN. PAINT BOTH TOP SURFACES WITH FLAT BLACK PAINT.

13"

25"

2-1/2"

AIR FLOW DIVIDER PANELS.

56"

2-1/2" TEMPORARY SPACER BLOCKS

25"

PANEL WILL BE MOUNTED HERE

NOTE 4" GAP AT FOOT

USING 2-1/2" SPACER BLOCKS, GLUE AND PIN (WITH #8 FINISHING NAILS) DIVIDER PANELS IN PLACE. REMOVE SPACERS. SEAL ALL INTERNAL SEAMS AND RAW PANEL EDGES WITH ALUMINUM TAPE.

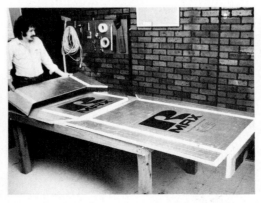

SOLAR COLLECTORS & WATER HEATERS

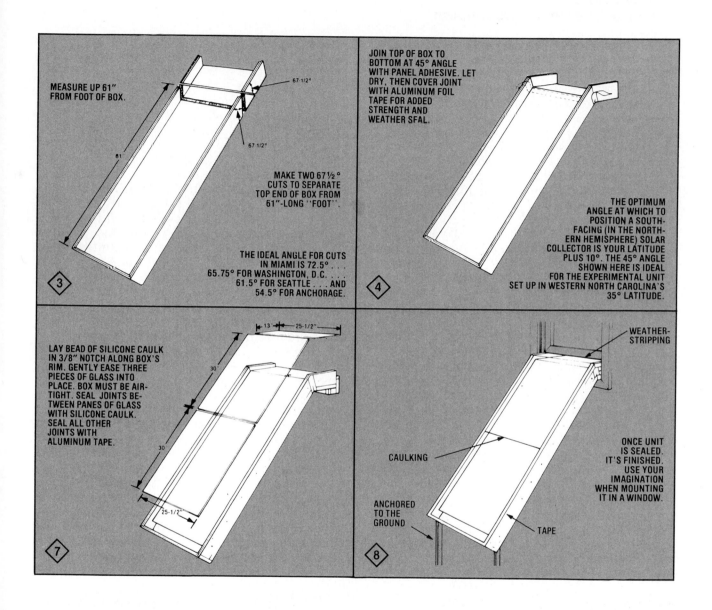

3 MEASURE UP 61" FROM FOOT OF BOX.

67·1/2°

67 1/2°

61

MAKE TWO 67½° CUTS TO SEPARATE TOP END OF BOX FROM 61"-LONG ''FOOT''.

THE IDEAL ANGLE FOR CUTS IN MIAMI IS 72.5° . . . 65.75° FOR WASHINGTON, D.C. . . . 61.5° FOR SEATTLE . . . . AND 54.5° FOR ANCHORAGE.

4 JOIN TOP OF BOX TO BOTTOM AT 45° ANGLE WITH PANEL ADHESIVE. LET DRY, THEN COVER JOINT WITH ALUMINUM FOIL TAPE FOR ADDED STRENGTH AND WEATHER SEAL.

THE OPTIMUM ANGLE AT WHICH TO POSITION A SOUTH-FACING (IN THE NORTH-ERN HEMISPHERE) SOLAR COLLECTOR IS YOUR LATITUDE PLUS 10°. THE 45° ANGLE SHOWN HERE IS IDEAL FOR THE EXPERIMENTAL UNIT SET UP IN WESTERN NORTH CAROLINA'S 35° LATITUDE.

7 LAY BEAD OF SILICONE CAULK IN 3/8" NOTCH ALONG BOX'S RIM. GENTLY EASE THREE PIECES OF GLASS INTO PLACE. BOX MUST BE AIR-TIGHT. SEAL JOINTS BE-TWEEN PANES OF GLASS WITH SILICONE CAULK. SEAL ALL OTHER JOINTS WITH ALUMINUM TAPE.

13" 25-1/2"

30"

30"

25-1/2"

8 WEATHER-STRIPPING

CAULKING

ONCE UNIT IS SEALED, IT'S FINISHED. USE YOUR IMAGINATION WHEN MOUNTING IT IN A WINDOW.

ANCHORED TO THE GROUND

TAPE

The rigid polyisocyanurate insulation board that forms the frame and divider of the heat grabber panel is an easy material with which to work.

STAR POWER

The Greedy Heat Grabber— A Solar Room Heater That Can't Help But Work (continued)

The basic foil-faced polyisocyanurate sheet *does* have a slight disadvantage, though. Its aluminum foil surfaces can be punctured relatively easily by anyone intent on doing just that. There are, however, at least two remedies for this problem: Substitute sheets which are faced on one side with a much heavier layer of aluminum, or use the regular boards and protect the sides, and bottom, and back of the finished collector with a casing of scrap lumber. The second alternative is less expensive than the first, but neither course of action should be necessary unless you live in a high-vandalism area.

Rigid polyisocyanurate board is so easy to work with that you won't need any saws, hammers, or other "conventional" carpentry tools to build this solar collector. The heat grabber here, in fact, was constructed with little more than a protractor, tape measure, paint brush, and two little hand-built knives.

Each knife is basically a block of 1″ X 2-1/2″ hardwood cut to fit the hand comfortably. The pieces of wood were then slotted and rigged with 10-32 bolts and wing nuts to grip Stanley 1992-5 utility knife blades at either a 45° (for "V" cuts) or a 90° (for square cuts) angle to the blocks' faces.

All cuts on the insulating board used in the collector were made straight and accurate by sliding one or the other of the two knives along a board or other straightedge that had been clamped to the rigid sheets of foam. For "V" cuts, the blade in the 45° knife was set to slice only to within about 1/32″ of the aluminum facing on the "far" side of the sheet (*not* all the way through either the facing *or* the foam). Since the foam varies slightly in thickness, this setting— for the most part—kept the blade from cutting too deeply. Two such cuts (with the straightedge reset for the second) were necessary for the completion of each "V".

And if you don't want to make "V" cuts and fold up the box of your solar collector? Then just build your "heat grabber" from separate pieces of foam board (all made with right angle cuts), peel back the aluminum skin from the butted face of each joint, and glue the sections—foam to foam—together.

The ideal angle at which to position a south-facing solar collector in the Northern Hemisphere (or a north-facing collector in the Southern Hemisphere) is your latitude plus 10°. The heat grabber in the diagrams was built in North Carolina at a site 35° north of the equator. The placement angle is therefore 45°. Take this into consideration when building a collector to be used elsewhere.

Remember that all the dimensions given in the plans are for a collector specifically tailored to fit the windows in one particular house. For a window that is wider or narrower, build the heat grabber accordingly. Also, it's unnecessary to keep the upper and lower air chambers in the collector *exactly* as deep as shown here. A variation of a half-inch or more is fine. As a matter of fact, it's awfully hard to keep this little Btu-grabber from working, as long as its passages are deep enough for air to circulate through them.

A word of caution: Although the single-strength glass specified is the same type of glass currently in use in millions of storm doors and windows, it *can* break and possibly will cut you or a child if, for any reason, either of you falls into it. Take whatever measures you deem necessary so that such an accident never happens.

LIST OF MATERIALS

Quantity	Material
1 sheet	1″ X 4′ X 8′ foil-faced rigid polyisocyanurate insulation
1/2 sheet +	3/4″ X 4′ X 8′ foil-faced rigid polyisocyanurate insulation
1 tube	Panel adhesive
1/2 tube	Silicone caulking compound
16	No. 8 finishing nails
3 pieces	Single-strength glass cut to fit
1/4 roll	All-metal aluminum foil duct tape
1 quart	Flat-black paint

SOLAR COLLECTORS & WATER HEATERS

Another Btu-Catcher to Last a Lifetime

While rigid polyisocyanurate foam board is probably the best material from which to fabricate a heat grabber (as detailed on the preceding pages), you can also frame the unit with wood when the former substance is not readily available. While the actual operation of these two heat grabbers is basically the same, a wooden collector will be heavier, will take considerably longer to construct, and may not work quite as well as the one made of insulative sheathing. But the good news is that a unit built from this new design [1] includes some refinements that improve its efficiency, [2] will most likely cost you only a few dollars per square foot (and even less than that if you use secondhand lumber), [3] is

more rugged, and [4] calls for materials you should have no trouble finding.

Start construction of the frame by cutting two 87"-long and two 27"-long 1 X 8's. Rout a 3/16" X 3/8" groove (3/4 inch in from one edge) down the length of the two long boards and one of the short ones as shown. Measure down 14 inches from the "top" ends of both long boards and scribe two marks (one "ahead of" and one "behind" the 14" mark) on each board. Then cut out this wedge of wood, so that two angles are left (see the drawing on page 20).

Rabbet all the frame's corners, cut another rabbet into the top inside edge of the frame (all the way around including the short pieces, to hold the glass that will eventually cover the box), cut another 3/16" X 3/8" groove in each 14" length of framing as shown, and make a half-lap joint in both sides of the frame where the angles will join. This is a good time to rip a 27" length of 1 X 8 into two pieces: One 2-1/4" and the other 4-3/4" wide. Trim and rabbet both boards to the dimensions shown.

Cut two 3/4" X 4" X 4" squares from the remains of your 1 X 8 stock (which actually measures only 3/4" thick and 7-1/4" wide), then seal all the surfaces of the framing wood with a good, clear preservative.

Saw a 26-1/4"-wide strip from a 3/16" sheet of paneling and cut that strip into three sections: 69-1/4", 12-3/4", and 7-3/4" long. Assemble the half-lap joints on the sides of the frame, trim the "point" at the top of each joint, and then slip the 26-1/4" X 69-1/4" piece of paneling—face up— into its groove along the frame's bottom edge. Fasten the "foot" to the frame's two sides with No. 6 X 1-3/4" wood screws.

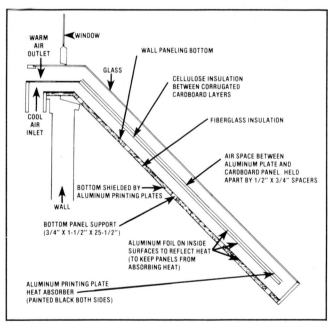

WARM AIR OUTLET
WINDOW
WALL PANELING BOTTOM
GLASS
CELLULOSE INSULATION BETWEEN CORRUGATED CARDBOARD LAYERS
COOL AIR INLET
FIBERGLASS INSULATION
AIR SPACE BETWEEN ALUMINUM PLATE AND CARDBOARD PANEL. HELD APART BY 1/2" X 3/4" SPACERS
BOTTOM SHIELDED BY ALUMINUM PRINTING PLATES
WALL
BOTTOM PANEL SUPPORT (3/4" X 1-1/2" X 25-1/2")
ALUMINUM FOIL ON INSIDE SURFACES TO REFLECT HEAT (TO KEEP PANELS FROM ABSORBING HEAT)
ALUMINUM PRINTING PLATE HEAT ABSORBER (PAINTED BLACK BOTH SIDES)

STAR POWER

Another Btu-Catcher
to Last a Lifetime (continued)

Slip the 7-3/4" X 26-1/4" piece of paneling into the lower groove in the upper (14"-long) section of the frame. Attach the 3/4" X 7-1/4" X 27" end piece to the lower 3-1/4" of the upper frame's sides. Take care to mount the board so its top edge is flush with the bottom of the grooves that you've made on the inside of that part of the frame.

Fasten the 3/4" X 4" X 4" blocks to the ends of the 3/4" X 7-1/4" X 27" board and then attach the 3/4" X 4-3/4" X 27" piece of framing to the blocks. Tack the 7-3/4" X 26-1/4" piece of paneling that's already in the frame to this board with small 1" finishing nails.

Slide the remaining (12-3/4" X 26-1/4") piece of paneling into the remaining set of grooves cut into the upper section of the collector's frame and tack it to the 3/4" X 7-1/4" X 27" end board. Then cut enough aluminum foil to cover the top surface of the 12-3/4" X 26-1/5" and the 26-1/4" X 69-1/4" sections of paneling that you've mounted in the frame. Coat the tops of both panels with water glass (sodium silicate, available from a pharmacy), carefully lay the precut pieces of foil on the boards as you would wallpaper (shiniest side up), and smooth out any air bubbles with a small block of wood. At this point you can complete the frame by fastening the 3/4" X 2-1/4" X 26-1/4" board (use small finishing nails) between the ends of the 14" side panels, at the top.

Turn the frame over, cut five 3/4" X 1-1/2" X 25-1/2" support strips from your 1 X 8 leftovers, space the strips about 17 inches apart with their 1-1/2" sides flat against the paneling that's already in the frame, drill through the sides of the collector into the ends of the strips, and fasten them in place with wood screws.

Trim fiberglass batting to fit between the support strips and use it to cover all of the remaining exposed undersurface of the paneling. Then—with the collector still upside down—cut enough aluminum printing plates (you can get them used in newspaper size from most printers for as little as $1.00 each) to cover the entire bottom of the heat grabber, allowing for 1" overlaps at the seams and for the 90° bend where the sheeting is fastened to the lip at the collector's top, and staple the plates to the edge of the frame and the cross-supports.

Turn the heat grabber back over and begin construction of the two-part divider by cutting five 1/2" X 3/4" X 67-1/2" strips and four 1/2" X 3/4" X 24" strips of wood from the remainder of your 1 X 8 stock. Then nail two of the 1/2" X 3/4" X 67-1/2" and all four of the 1/2" X 3/4" X 24" pieces of wood together into a 1/2" X 25-1/2" X 67-1/2" "ladder".

Trim two sheets of corrugated cardboard to the same outside dimensions as the ladder framework (25-1/2" X 67-1/2"), staple one of the sheets to one side of the frame, turn the assembly over, and glue the 1/2"-thick tubing spacers to the cardboard's inside surface.

Fill this newly formed "tray" with an ample amount of cellulose insulation (or fiberglass batting), lay a bead of strong glue on the exposed surfaces of the 1/2"-thick spacers and framing, and staple on the other sheet of cardboard to form an insulation-and-spacer "sandwich".

Cut enough aluminum foil to cover both pieces of cardboard. Then coat one side of the divider with water glass, carefully cover it with foil, turn the divider over, and cover its other side the same way.

SOLAR COLLECTORS & WATER HEATERS

Next, position the three remaining 1/2″ X 3/4″ X 67-1/2″ strips of wood on top of the foil-covered divider—3/4″ sides down and spaced as shown (7-3/4″ apart)—and nail them to the divider's frame and crossmembers. Then cut enough of the aluminum printing plate material to cover the 25-1/2″ X 67-1/2″ area (allow enough for overlap at the seams), cover it well on both sides with a coat or two of high-temperature flat black paint, and staple the aluminum sheeting to the three supports.

Position the completed divider in the collector's frame—aluminum plate up—so that it's butted up against the foil-covered paneling in the heat grabber's upper section (this will leave about a 2-3/4″ space between the divider's bottom end and the frame's lower end board). Hold the divider up temporarily so that there's a 2-1/4″ space all the way up and down between its back surface and the foil-covered paneling underneath it. Then drill through the heat grabber's sides into the sides of the divider's framing and secure the divider in place with No. 6 X 3/4″ flathead wood screws.

Lay the three panes of 1/4″ single-strength glass into the groove around the top of the collector's frame, secure them in place with glazier's points, run a bead of silicone sealant along the two seams where the panes meet, and form a weathertight seal between the glass and the wooden frame with glazing compound.

Now just hook the "lip" of the heat grabber over the sill of a south-facing window, prop up the base of the panel, shut the window firmly on the collector's top . . . and sit back and enjoy the warmth that this room heater will instantly start pumping during sunny weather.

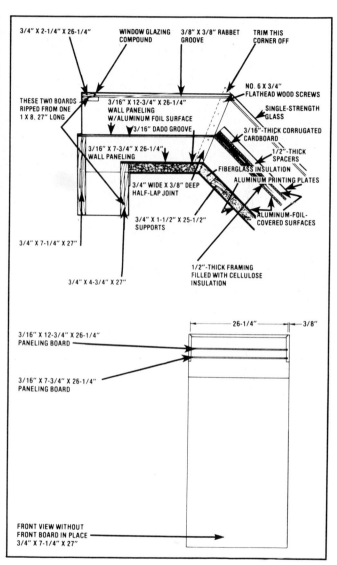

STAR POWER

Another Btu-Catcher
to Last a Lifetime, (Continued)

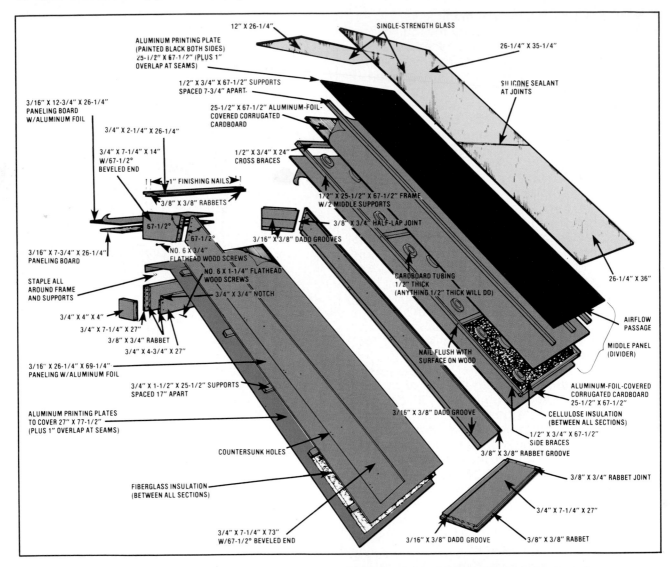

ALUMINUM PRINTING PLATE (PAINTED BLACK BOTH SIDES) 25-1/2" X 67-1/2" (PLUS 1" OVERLAP AT SEAMS)

12" X 26-1/4"

SINGLE-STRENGTH GLASS

26-1/4" X 35-1/4"

SILICONE SEALANT AT JOINTS

3/16" X 12-3/4" X 26-1/4" PANELING BOARD W/ALUMINUM FOIL

1/2" X 3/4" X 67-1/2" SUPPORTS SPACED 7-3/4" APART

25-1/2" X 67-1/2" ALUMINUM-FOIL-COVERED CORRUGATED CARDBOARD

3/4" X 2-1/4" X 26-1/4"

1/2" X 3/4" X 24" CROSS BRACES

3/4" X 7-1/4" X 14" W/67-1/2° BEVELED END

1" FINISHING NAILS

1/2" X 25-1/2" X 67-1/2" FRAME W/2 MIDDLE SUPPORTS

3/8" X 3/8" RABBETS

3/8" X 3/4" HALF-LAP JOINT

67-1/2°

3/16" X 3/8" DADO GROOVES

26-1/4" X 36"

3/16" X 7-3/4" X 26-1/4" PANELING BOARD

67-1/2°

NO. 6 X 3/4" FLATHEAD WOOD SCREWS

CARDBOARD TUBING 1/2" THICK (ANYTHING 1/2" THICK WILL DO)

STAPLE ALL AROUND FRAME AND SUPPORTS

NO. 6 X 1-1/4" FLATHEAD WOOD SCREWS

AIRFLOW PASSAGE

3/4" X 4" X 4"

3/4" X 3/4" NOTCH

MIDDLE PANEL (DIVIDER)

3/4" X 7-1/4" X 27"

3/8" X 3/4" RABBET

NAIL FLUSH WITH SURFACE ON WOOD

3/4" X 4-3/4" X 27"

3/16" X 26-1/4" X 69-1/4" PANELING W/ALUMINUM FOIL

ALUMINUM-FOIL-COVERED CORRUGATED CARDBOARD 25-1/2" X 67-1/2"

3/4" X 1-1/2" X 25-1/2" SUPPORTS SPACED 17" APART

CELLULOSE INSULATION (BETWEEN ALL SECTIONS)

ALUMINUM PRINTING PLATES TO COVER 27" X 77-1/2" (PLUS 1" OVERLAP AT SEAMS)

3/16" X 3/8" DADO GROOVE

1/2" X 3/4" X 67-1/2" SIDE BRACES

3/8" X 3/8" RABBET GROOVE

COUNTERSUNK HOLES

3/8" X 3/4" RABBET JOINT

FIBERGLASS INSULATION (BETWEEN ALL SECTIONS)

3/4" X 7-1/4" X 27"

3/4" X 7-1/4" X 73" W/67-1/2° BEVELED END

3/16" X 3/8" DADO GROOVE

3/8" X 3/8" RABBET

COLLECTOR SIZE: 15.2 SQUARE FEET

LIST OF MATERIALS

QUANTITY	MATERIALS
34 feet	1 X 8 pine (use cedar if available) cut into: (2) 8 ', (1) 7 ', (1) 6 ', and (1) 5 ' length(s)
1 piece	26-1/4" X 84" single-strength glass (cut into: 12", 35-1/4", and 36" lengths)
1 sheet	3/16" X 4' X 8' wall paneling
1 box	Glazier's points
1 can	Glazing compound
1 quart	Water glass or sodium silicate (available at drugstore)

SOLAR COLLECTORS & WATER HEATERS

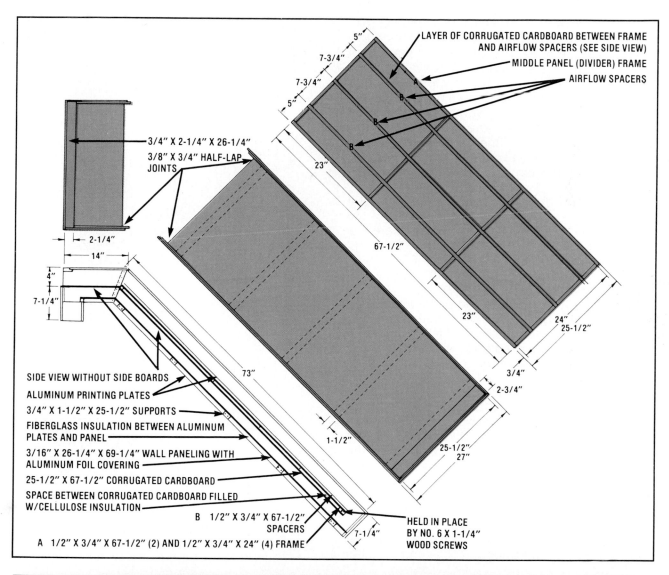

LAYER OF CORRUGATED CARDBOARD BETWEEN FRAME AND AIRFLOW SPACERS (SEE SIDE VIEW)

MIDDLE PANEL (DIVIDER) FRAME

AIRFLOW SPACERS

3/4" X 2-1/4" X 26-1/4"

3/8" X 3/4" HALF-LAP JOINTS

5"
7-3/4"
7-3/4"
5"
23"
67-1/2"
23"
24"
25-1/2"
3/4"
2-3/4"
25-1/2"
27"
1-1/2"
73"
2-1/4"
14"
4"
7-1/4"

SIDE VIEW WITHOUT SIDE BOARDS

ALUMINUM PRINTING PLATES

3/4" X 1-1/2" X 25-1/2" SUPPORTS

FIBERGLASS INSULATION BETWEEN ALUMINUM PLATES AND PANEL

3/16" X 26-1/4" X 69-1/4" WALL PANELING WITH ALUMINUM FOIL COVERING

25-1/2" X 67-1/2" CORRUGATED CARDBOARD

SPACE BETWEEN CORRUGATED CARDBOARD FILLED W/CELLULOSE INSULATION

B   1/2" X 3/4" X 67-1/2" SPACERS

A   1/2" X 3/4" X 67-1/2" (2) AND 1/2" X 3/4" X 24" (4) FRAME

7-1/4"

HELD IN PLACE BY NO. 6 X 1-1/4" WOOD SCREWS

LIST OF MATERIALS (continued)

QUANTITY	MATERIALS	QUANTITY	MATERIALS
8 pieces	Aluminum printing plates (newspaper size), available from many printing or newspaper shops	1 tank's supply	Fiberglass batting insulation (taken from discarded hot water heaters)
12	No. 6 X 3/4" flathead wood screws	1 box	18" X 38' heavy aluminum foil
54	No. 6 X 1-1/4" flathead wood screws	1 box	3/8" wood staples
2 sheets	25-1/2" X 67-1/2" corrugated cardboard (cut from discarded refrigerator boxes)	1 length	18" X 3" cardboard tubing
		1/2 pint	High-temperature flat black paint can
1 quart	Clear wood preservervative	1/10 tube	Silicone sealant
5 pounds	Cellulose insulation	34	Small finishing nails, 1" long
		1 bottle	Glue

STAR POWER

A "High-rise Heat Grabber" for Everyone

The preceding sections of this chapter discuss two ways of building a passive solar collector that can be made-to-measure for just about any ground-floor, double-hung window that faces south. The simple convection-cycle operation of one of these window-hung panels is explained, pointing out the many advantages to be gained from installing a homemade heat grabber. But suppose you're an apartment or high-rise dweller? It's a bit difficult to provide ground support for a solar panel when your view is from the fourteenth floor!

So here is a *high-rise* heat grabber that still provides a low-cost way to tap the energy of the sun. Not only can it be used anywhere there is a south-facing window that opens upward, but this modified version is also easy to handle, is virtually unbreakable, and is therefore quite safe to use, especially around children.

The secret is in building the collector surface of lightweight, nonreflective, energy-transmitting, and almost vandalproof plastic (the same fiberglass-reinforced material that's commonly used in modern greenhouses and commercially made solar collectors).

Just a shade less noteworthy is that the two sections of this heat-grabber are perpendicular to each other so the collector mounts flush with the outside wall and then "right-angles" into the room to be heated. This feature eliminates the need for a ground-floor installation (for that matter, the heater no longer needs any kind of lower support since the whole affair weighs only 14 pounds) and does away with any bothersome optimal-angle calculations.

But that's not all: The inexpensive solar heat pump now uses a thermostatically controlled fan

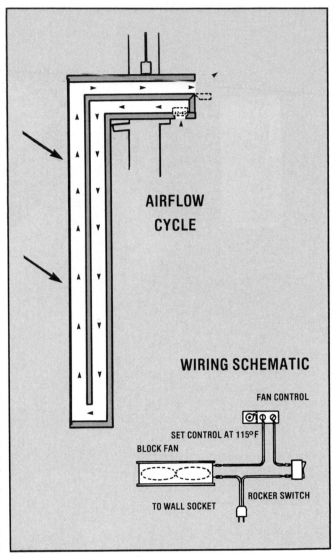

AIRFLOW CYCLE

WIRING SCHEMATIC

FAN CONTROL

SET CONTROL AT 115°F

BLOCK FAN

TO WALL SOCKET

ROCKER SWITCH

Mounted well above ground level in the south-facing window of an apartment, the high-rise heat grabber can cut heating costs for city-bound renters.

SOLAR COLLECTORS & WATER HEATERS

NECK AND BODY CUTTING DIAGRAM
1" X 4' X 8' SHEETS

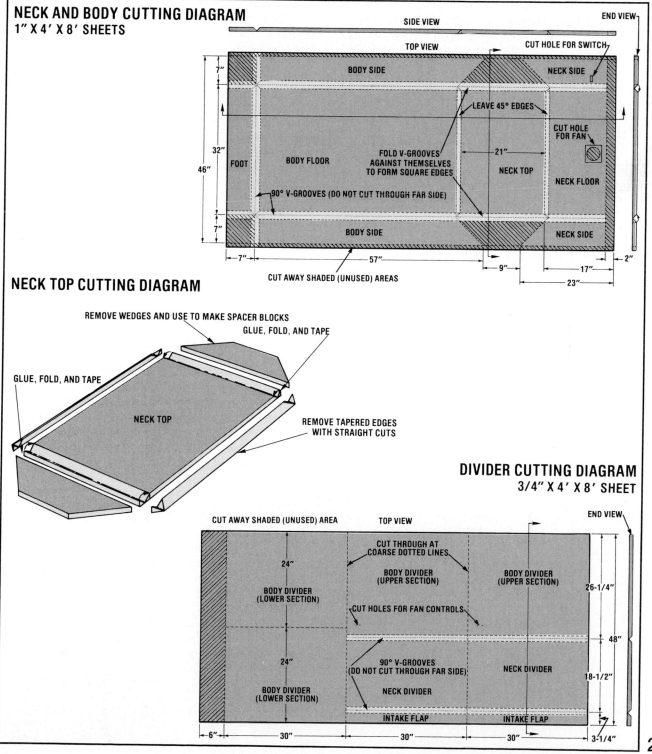

SIDE VIEW

END VIEW

TOP VIEW

CUT HOLE FOR SWITCH

7"

BODY SIDE

NECK SIDE

LEAVE 45° EDGES

CUT HOLE FOR FAN

32"

FOOT

BODY FLOOR

FOLD V-GROOVES AGAINST THEMSELVES TO FORM SQUARE EDGES

21"

NECK TOP

NECK FLOOR

46"

90° V-GROOVES (DO NOT CUT THROUGH FAR SIDE)

7"

BODY SIDE

NECK SIDE

7"

57"

9"

17"

2"

CUT AWAY SHADED (UNUSED) AREAS

23"

NECK TOP CUTTING DIAGRAM

REMOVE WEDGES AND USE TO MAKE SPACER BLOCKS

GLUE, FOLD, AND TAPE

GLUE, FOLD, AND TAPE

NECK TOP

REMOVE TAPERED EDGES WITH STRAIGHT CUTS

DIVIDER CUTTING DIAGRAM
3/4" X 4' X 8' SHEET

END VIEW

CUT AWAY SHADED (UNUSED) AREA

TOP VIEW

CUT THROUGH AT COARSE DOTTED LINES

24"

BODY DIVIDER (UPPER SECTION)

BODY DIVIDER (UPPER SECTION)

26-1/4"

BODY DIVIDER (LOWER SECTION)

CUT HOLES FOR FAN CONTROLS

48"

24"

90° V-GROOVES (DO NOT CUT THROUGH FAR SIDE)

NECK DIVIDER

18-1/2"

BODY DIVIDER (LOWER SECTION)

NECK DIVIDER

INTAKE FLAP

INTAKE FLAP

6"

30"

30"

30"

3-1/4"

23

STAR POWER

A "High-rise Heat Grabber" for Everyone (continued)

to draw cool room air into the unit and force sun-heated air back to the house. On days when large volumes of warm currents aren't needed, the blower can be shut off, and the collector will simply cycle passively, courtesy of the sun.

In all fairness, however, there is a trade-off involved in the switch from a 45° (more or less) configuration to one of 90°, and this is because the window furnace must give up as much as 30% of its efficiency for the sake of its "against the wall" mounting convenience. Fortunately, a sacrifice of this degree will occur only at the beginning and end of each winter season—when the sun still crosses the sky at a high angle—and will probably be reduced to an average of 10% or so during the bulk of the heating period.

The only tools required for assembling a high-rise heat grabber—other than the homemade knives detailed in the first section of this chapter—are a tape measure, a caulking gun, a screwdriver, and perhaps a sharp kitchen knife.

By following the cutting diagrams, you'll be able to carve up enough sections from two sheets of insulation board to make two complete heat grabbers with a minimum of leftover waste, provided that your window opening—from track to track—is 32" in width. You can make your solar furnace a maximum of 34" wide by using the entire breadth of a 48" rigid foam sheet for the base and sides, or conversely, you can reduce the heater's width to fit a narrow window by cutting away the appropriate amount of material.

Actually, as long as you approximate the specifications for the 2-1/2" chamber depths and the 3" divider-to-foot space, it'll be difficult to keep this convective collector from working. And should you want to install the unit in an opening larger than 34", simply cut some of your leftover board into spacers that'll fit snugly between the sides of the grabber and the tracks in the window frame. In any case, once the sash is drawn tightly against the top of the spacers and the collector, the installation joint should be sealed all around with duct tape to discourage drafts.

One note of caution, however: If you plan to use your lightweight solar collector on a high-rise building or at a similar location where buffeting winds might present a problem, you might well consider "double-checking" the device by running a sturdy wire across its lower face (or better yet, installing some screw eyes through its sides with body washers before putting the glazing on), and then fastening the box to something convenient and sturdy—such as a window frame or a ventilation grille—to secure it against especially strong breezes.

The accompanying diagrams will show you everything else you need to know to build this economical city-dweller's solar heater. So why not shake those what-can-I-do urban blues? Build yourself a pair of high-rise heat grabbers (or, better still, share the project—and the rewards—with a friend or neighbor), and grab your *own* fair share of the sun!

LIST OF MATERIALS	
Quantity	**Item**
2 sheets	1" X 4' X 8' foil-faced rigid foam insulation board
1 sheet	3/4" X 4' X 8' foil-faced rigid foam insulation board
2 pieces	.025" X 3' X 5' fiberglass-reinforced plastic glazing aluminum foil duct tape
2 rolls	Aluminum foil duct tape
2 cans	13-ounce high-temperature flat black paint
2 units	120-VAC, 70-CFM block fan
2 units	120-VAC, 6-amp rocker switch
1 tube	Paneling adhesive
1 tube	Silicone adhesive
2 units	120-VAC, 6-amp warm air differential fan control
2 units	Nine-foot light-duty extension cords
4 units	1/4" body washers
4 units	3/16" X 1-3/4" roundhead bolts with nuts

COLLECTOR SIZE: 14 square feet

NOTE: Because of the manner in which many of the required materials are packaged and sold, it's more economical to make *two* heat grabbers than to make one, thereby avoiding waste. Also, many of the materials listed might be found in your workshop or scrap pile, thus significantly reducing each heater's overall cost.

SOLAR COLLECTORS & WATER HEATERS

"Sun Bulbs"—
A Fluorescent Tube Furnace

New applications for "used up" materials that our society discards daily as scrap are limited only by the human imagination and ingenuity. Fluorescent tubes, for instance, that are cast aside when they no longer serve their original purpose of providing light in commercial or public buildings or in our homes can be resurrected to function as the central part of a built-from-scrap solar collector and heater.

Start this project by collecting a healthy supply of burned-out 96"-long fluorescent lights, available for free from offices, factories, stores, shopping centers, dumps, and other similar sources. Twenty-nine tubes are required for the collector described here, but it's a good idea to pick up a few extras in the beginning, to allow for breakage.

Because the coating on the insides of fluorescent tubes contains beryllium and other chemicals that can cause real problems if they get into cuts or scrapes, heavy gloves should always be worn when handling the bulbs. Also, since there is a possibility that fumes containing these chemicals could be released when a tube is cut

While somewhat larger than a conventional clothes dryer, the fluorescent tube collector captures the sun's energy to produce heat rather than electricity. Variations and modifications of the basic unit are easily made to suit a person's needs.

25

STAR POWER

"Sun Bulbs"—
A Fluorescent Tube Furnace (continued)

or broken, it's best to work with the cylinders outdoors.

Puncture the metal or plastic ends of the long lights to relieve the vacuum inside, then use a fluorescent tube cutter (built as described in the next section) to remove one end from each tube. Now, pour some sharp sandblasting sand into the cylinders and tip them back and forth repeatedly until all the coating has been removed from their insides. Then cut enough off the remaining end of each tube to leave clean, transparent glass cylinders, 85" long and open on both ends. Finally, paint all 29 of the glass pipes flat black inside and out. Use a cotton pad, pulled through each tube, to spread the paint deep inside, and spray or brush paint on the outsides.

Next, rout a 1/8" X 3/8"-deep groove in the sides and ends of the collector box 3/4 inch from the bottom edges, rabbet the corners of the end boards, trim the 1/8" paneling to fit, and fasten the whole box together with wood screws. Rip one of the remaining 1 X 6's into four pieces and screw them to the frame to support the underside of the paneling. Using water glass (sodium silicate) as glue, attach aluminum foil (shiny side up) to the topside.

Cut the two 1 X 6 tube holders down to a width of 4-3/4", then draw a line down the middle of both boards and—starting 9/16 inch in from one end—mark the 29 center points (1-9/16" on centers) for the mounting of the modified fluorescent tubes. Next, drill out all the tube mounting holes with a 1-1/2" hole saw and cut the two boards apart on their centerlines. One half of each holder should then be securely fastened inside the collector box (5-1/8 inches in from each end). Lay a bead of silicone sealant along the surface of the holes in the mounted half-holders and set each cut-and-painted glass tube in place. Spread sealant on the holes in the other halves of the holders and attach the scalloped pieces of wood to the sides of the box with wood screws.

A length of T-bar (ordinarily used to support a suspended ceiling) is added next as bridging between the centers of the tube holders to hold up the 5-mil UV-X film that will cover the collector. Mortise a groove and cut a slot into the middle of each holder so the top of the T-bar will rest flush with the tops of the wood pieces.

Now's the time (before you close your collector and make it too difficult to work on) to cut inlet and outlet air holes into one side of the box with a 3" hole saw. Smooth the edges of the holes with sandpaper, clean the sawdust out of the box, and slap a good coat of wood sealer onto all the exposed wood surfaces.

Then stretch and staple the UV-X film to the box's top. The plastic covering should be further secured and protected—all the way around the face of the collector—with 3/4" X 3/4" outside corner molding. To keep air from blowing over the tube supports and past, rather than through, the painted collector tubes, lengths of 1/8" X 3/4" screen molding should be cut to fit, laid down across the UV-X film, and securely screwed in place along the tops of the tube holders.

That's it. If you hook a small (100- to 200-cubic-feet-per-minute) blower to your new collector's inlet and aim the energy catcher at the sun, you'll find that on a clear day the unit will give you up to 6,250 Btu per hour, which is enough to heat a room or to dry your clothes.

SOLAR COLLECTORS & WATER HEATERS

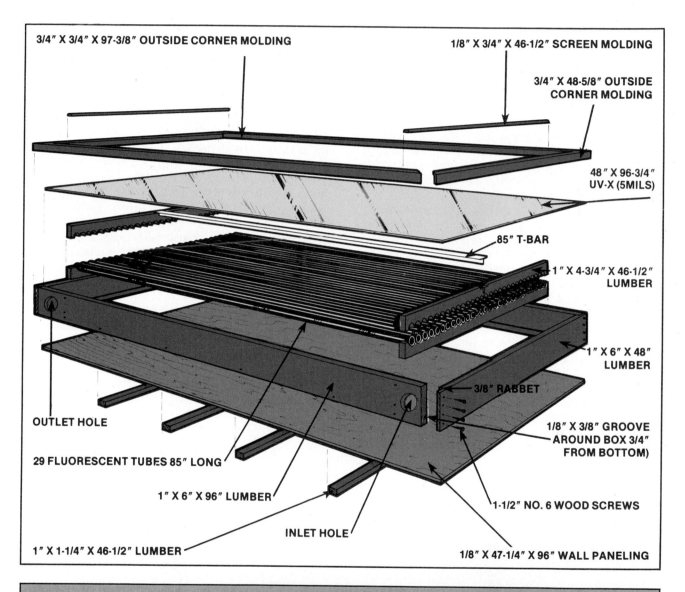

3/4" X 3/4" X 97-3/8" OUTSIDE CORNER MOLDING

1/8" X 3/4" X 46-1/2" SCREEN MOLDING

3/4" X 48-5/8" OUTSIDE CORNER MOLDING

48" X 96-3/4" UV-X (5MILS)

85" T-BAR

1" X 4-3/4" X 46-1/2" LUMBER

1" X 6" X 48" LUMBER

3/8" RABBET

OUTLET HOLE

29 FLUORESCENT TUBES 85" LONG

1/8" X 3/8" GROOVE AROUND BOX 3/4" FROM BOTTOM)

1" X 6" X 96" LUMBER

1-1/2" NO. 6 WOOD SCREWS

INLET HOLE

1" X 1-1/4" X 46-1/2" LUMBER

1/8" X 47-1/4" X 96" WALL PANELING

MATERIALS LIST

(29) Used fluorescent tubes, unbroken
(1) 48" X 96" sheet of wall paneling
(2) 1" X 6" X 48" stock lumber
(2) 1" X 6" X 96" stock lumber
(2) 1" X 6" X 46-1/2" stock lumber
(1) 1" X 6" X 46-1/2" stock lumber (cut into four pieces, each 1-1/4" wide)

(2) 3/4" X 48-5/8" outside corner molding
(2) 3/4" X 97-3/8" outside corner molding
(2) 3/4" X 46-1/2" screen molding
(1) 85" suspended ceiling T-bar
(1) 48" X 96-3/4" clear weather-resistant-type UV-X polyester film, 5 mils thick

(40) 1-1/2" No. 6 wood screws
(50) 3/4" No. 4 wood screws
(1) Tube of silicone sealant
(1) Quart can flat-black paint
(1) Pint can of wood sealer
(1) Quart can of sodium silicate nails, glue, and aluminum foil
(1) 100–200-cfm blower

STAR POWER

How to Cut the Tubes for a Fluorescent Furnace

Besides the challenge of adapting simple—and in many instances, scrap—materials into useful devices that will harness natural energy, specialized tools are often required to work with the "unusual" objects that are incorporated into homemade solar collectors and other energy-saving inventions.

As an example, cutting the ends off discarded fluorescent tubes used as the collection unit of a solar heater presents a special kind of problem, but one that can be easily solved by making a cutter to suit the task.

The principle of the cutter discussed here is quite elementary: A stainless steel wire, which can be heated electrically, encircles the glass fluorescent tube. When the current is on, the wire heats and melts into the glass. When the power is turned off, the glowing wire immediately cools. The glass then cools unevenly and breaks apart on the melted "score line".

A cutter can be assembled in about an hour, using a hammer, a drill, a handsaw, some assorted bits, a center punch, a tap and die set, a wood chisel, a hacksaw, a pair of wire cutters, and a screwdriver.

Start by cutting a 1-1/4" X 7" wooden dowel exactly in half lengthwise. Then temporarily nail a short scrap of wood to the rounded side of each dowel half and clamp these pieces—first one then the other—in a vise so that the flat side of each half faces up.

With a handsaw, cut a 1/8"-deep lengthwise groove down the center of each half-dowel's flat side. Then, using a wood chisel, carefully open one end of each of these grooves into a wider and deeper slot that measures 5/16" wide, 3/8" deep, and 1-1/4" long.

Drill four 3/16" holes through the dowel halves (one hole 1/2 inch in from each end of each half). And use a 3/8" bit to bore two holes (each about 1/4" deep but not all the way through the wood) into the flat faces of the half-dowels about three inches from their "unslotted" ends and to one side of the center grooves.

Now, drill two 5/32" holes—one 3/16 inch from one end and the other 3/8 inch from the opposite end—through each of the two 5/16" X 5/16" X 1-1/4" square steel rods. Cut threads into each of these four holes with a 10-32 tap. Then put the square rods—one at a time—into the vise with their "short ends" out (the ends with the holes drilled through 3/16" back) and bore a 1/16" hole lengthwise into each block 3/32 inch from the edge, until the drill bit pierces through into the first tapped hole.

With this done, stick a length of .034" stainless steel wire into the 1/16" hole and thread a 10-32 machine screw into the end of the tapped hole closest to the wire until the screw touches the wire's edge. Cut the rest of the screw off flush with the side of the square rod, and then lock the sawed-off "stop" that is left in the hole firmly in place by nicking the cutoff stub with a center punch where it meets the face of the square rod.

Line the two wooden handle halves up with one another, tape them together, mark and drill the four 3/32" hinge-mounting holes in the handle's butt, screw the hinge firmly in place, and remove the tape.

Then take a length of good lamp cord, separate one end of its two insulated strands for about nine inches, and peel a half inch or so of insulation from the end of each strand. One of these

SOLAR COLLECTORS & WATER HEATERS

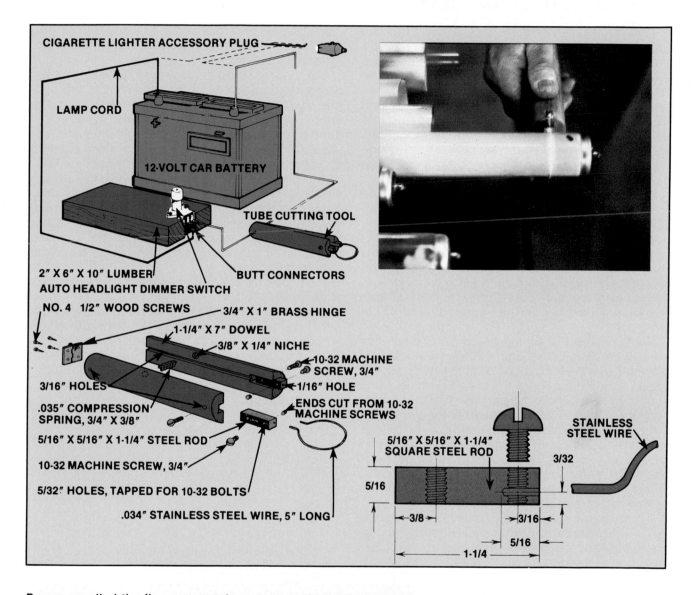

CIGARETTE LIGHTER ACCESSORY PLUG

LAMP CORD

12-VOLT CAR BATTERY

TUBE CUTTING TOOL

2" X 6" X 10" LUMBER

BUTT CONNECTORS

AUTO HEADLIGHT DIMMER SWITCH

NO. 4   1/2" WOOD SCREWS

3/4" X 1" BRASS HINGE

1-1/4" X 7" DOWEL

3/8" X 1/4" NICHE

10-32 MACHINE SCREW, 3/4"

3/16" HOLES

1/16" HOLE

.035" COMPRESSION SPRING, 3/4" X 3/8"

ENDS CUT FROM 10-32 MACHINE SCREWS

5/16" X 5/16" X 1-1/4" STEEL ROD

10-32 MACHINE SCREW, 3/4"

5/32" HOLES, TAPPED FOR 10-32 BOLTS

.034" STAINLESS STEEL WIRE, 5" LONG

5/16" X 5/16" X 1-1/4" SQUARE STEEL ROD

STAINLESS STEEL WIRE

3/32

5/16

3/8

3/16

5/16

1-1/4

Power supplied the fluorescent tube cutter reaches the business end of the handle through wires running up the tool's center. The stainless steel loop at the end cleanly cuts the ends from the bulbs.

STAR POWER

How to Cut the Tubes
for a Fluorescent Furnace (continued)

separated and peeled wires should then be threaded "in" through the 3/16″ hole on the butt end of each half of the tube cutter's wooden handle and run through the groove sawed into the flat face of the handle half so that its bare end can be seated under—and directly in contact with—the square steel block which mounts into the "business end" of each half of the handle. Next, slip the two steel blocks into position ("stopped" sides facing each other) and secure each one in place with a 10-32 machine screw.

Note that the steel blocks—and all metal hardware in each assembly—are recessed into the wood so that neither block or any of its components can touch the other and create a short circuit. Electrical current can travel down only one strand of the lamp cord to the steel block on its end, through a loop of stainless steel wire connecting the first steel block to the second, and then back through the second strand of the lamp cord.

Finally, take the 5″ length of .034″ stainless steel wire. Stick one end of the wire into the 1/16″ hole in the end of one of the cutter's square steel blocks and secure the wire tightly in place with a 10-32 machine screw. Then form the wire into a circle and clamp its other end into the other square block so that the ends of the handle will never quite touch when the loop is squeezed around a fluorescent tube. When this adjustment has been made, separate the handle halves and insert the small compression spring into the holes that were drilled for it.

At this point you're ready to hook up your cutter to a 12-volt car battery or an automobile cigarette-lighter tap. In any case, the unit's on/off switch is nothing more than an automobile headlight dimmer switch, available from any wrecking yard for pennies or from an auto supply house for little expense.

Cut a notch, if necessary, into the end of a 10″ piece of scrap 2 X 6 lumber and mount the dimmer switch to the board. Connect one of the lamp cord conductors coming from the cutter handle to one pole of the power source and, using a butt connector, attach the cord's other wire to one of the terminal lugs of the dimmer switch.

Then run a length of wire from the dimmer switch's other terminal (use another butt connector) to the remaining pole on the power supply. That's it. You're ready to cut fluorescent tubes, bottles, jars, and even water glasses (though the larger objects will need a longer stainless steel wire loop).

Just hook up your power source (alligator clips work fine), fit the loop of stainless steel wire around the glass tube you want to cut, and depress the foot switch. When the wire glows red hot, depress the switch again, and the cylinder of glass will be cut to order. The loop of stainless, of course, will break from time to time (due to its constant expansion and contraction), but it's easy to replace.

LIST OF MATERIALS

- (1) 1-1/4″ X 7″ wooden dowel
- (2) 5/16″ X 5/16″ X 1-1/4″ square steel rods
- (1) 5″ length of .034 stainless steel wire
- (4) 10-32 machine screws, 3/4″
- (1) 3/4″ X 1″ brass hinge
- (4) No. 4 X 1/2″ wood screws
- (1) 60″ length of lamp cord
- (1) .035 compression spring, 3/8″ X 3/4″

- (1) 12-volt car battery or (1) cigarette lighter adapter plug
- (1) Automobile headlight dimmer switch
- (2) No. 6 X 1/2″ wood screws
- (1) 10″ scrap of 2 X 6
- (2) Butt connectors for 14 to 16 gauge wire
- (2) Large alligator clips

SOLAR COLLECTORS & WATER HEATERS

A Crafty
Corrugated Collector

Harnessing the sun's energy for home heating can be done in many ways, and solar collectors designed to warm the air in a dwelling can take several forms and incorporate a variety of materials. Using sheets of corrugated fiberglass for glazing, insulating board, and 2 X 4's, a person can easily build a low-cost, blower-assisted, 96-square-foot collector. When attached to the south-facing wall of a home, it can deliver as many as 19,000 Btu per hour at no expense other than the minimal amount for the electricity required to run the small fan motor. This convective collector can be ready to hang in just a few hours, even if one is new to the carpentry game.

Start the construction of this sun-powered heater by locating the studs inside your frame wall and scouting for any plumbing or wiring which might get in the way of the collector's intake or exhaust vents. Once you've mapped out a suitable 8' X 12' location, just countersink holes every two feet in the 2 X 4's that will be the collector's frame. Make the holes about two inches deep (as shown in the illustration). Then fasten the parts to the wall with 3" nails.

Next, you can add the collector's internal baffles by countersinking and nailing *these* 2 X 4's to the wall just as you did the frame pieces. The spacing shown in the diagram is critical since it maintains the correct airflow through the system. Because they're 24 inches apart, some baffle boards should align with the studs.

Now that the frame is complete, cut the 3" X 12" vent holes in your wall, line them with rigid polyisocyanurate insulating board, and seal the cracks with silicone caulking. If you trim a standard-sized sheet of the insulator carefully, there should be enough material left to completely cover the rest of the exposed wall space inside the frame. Make sure that the insulation fits tightly around the 2 X 4's and then seal every junction with a liberal amount of caulking. When this step is completed, paint the inside of the collector flat black.

Once the paint is dry, lay each piece of precut ripple board (available from lumber or hardware stores) on an appropriate frame member, and secure them with three nails each. The translucent corrugated fiberglass fits right over the wood, and after you've drilled a hole that's three sizes smaller than the diameter of the rubber-sealed roofing nails (which secure the covering) through every third bump in both the wood *and* the glass, you can caulk the surfaces and affix the light-admitting material.

Finally, tack and carefully apply silicone-seal to a 12-foot strip of "drip edge" along the top and bottom of the collector to seal the fiberglass and to protect the wood from rain. Then cut a 2" X 8" vent in the vertical 2 X 4 next to the exhaust duct to allow the collector to breathe during extremely warm spells. This hole, however, should be plugged when the heater's in use. In addition, people who live in a very windy climate may wish to add an extra layer of glazing for insulation. Any substantial clear plastic sheeting of at least 3-mil thickness is suitable for this purpose.

To activate your solar heater, just insert a blower (about two cubic-feet-per-minute capacity for each square foot of collector) in the intake duct, and let the sunshine do the rest. In moderate climates, two of these devices can warm your entire abode at such a low cost that the amount saved on heating bills should cover the investment in materials in less than a year.

Built quickly and easily against a south wall where it can absorb sunlight, this solar collector can provide a home with 19,000 Btu per hour of heat during the winter.

STAR POWER

A Crafty
Corrugated Collector (continued)

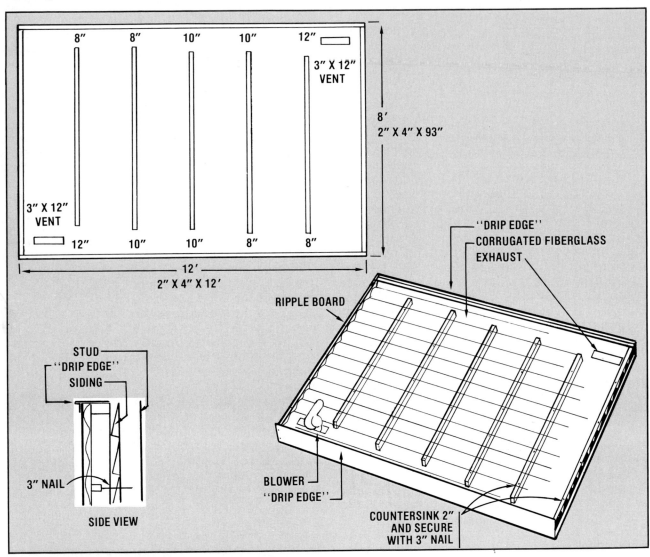

LIST OF MATERIALS

- (4) 2′ X 12′ sheets of corrugated fiberglass
- (3) 3/4″ X 4′ X 8′ sheets of rigid polyisocyanurate insulating board
- (9) 8′ strips of ripple board
- (2) 12′ strips of "drip edge"
- (1) quart of flat black paint
- (3) tubes of silicone caulking
- (7) 8′-long 2 X 4's
- (2) 12′-long 2 X 4's
  nails

SOLAR COLLECTORS & WATER HEATERS

A Crafty Corrugated Collector Control

The efficiency and moneysaving aspects of the solar corrugated collector so intrigued one homeowner that he decided to cover the entire south-facing wall—including the windows—of his house with that type of panel, increasing the heat gathering capabilities of the design dramatically. Since he was about to put roughly *200 square feet* of "solar furnace" into action, he enlarged the vent size to 6″ X 12″ and ran a 10″ duct through his attic to drop the warm air into the center of his home. With these modifications, he was able to use a blower with a flow rate of 400 cubic feet per minute.

Well, he did such a good job of reworking the collector that on sunny days the unit actually provided too much heat. So the inventive homeowner added thermostatic controls to the design so the blower would work automatically.

The control system works by placing one temperature sensor in the living area of the house and another inside the collector, then wiring the two sensors and the blower into a series circuit. Thus there must be both a need for heat and sufficient warmth on hand in the collector before the blower will switch on.

You can start the installation by lifting the upper panel of corrugated fiberglass to expose a point about a foot from the top and two "channels" over from the air intake. Drill a 1/2″ hole through the insulating board and the wall and run two wires through the opening. Then use screws to fasten a 70–160°F fan control to the collector surface and attach each of the wires to a pole on the control. The positive wire will go to a power source inside the house, and the neutral wire will connect to one terminal on a 115V thermostat located near the main thermostat for your heating

system. (Warning: Before you make the "hot" connection to your electrical system, be sure to shut off the circuit breaker for that section of your home's wiring to avoid shock.)

Once the hot-wire/control/thermostat circuit is complete, you can set the temperature scale on the collector's sensor to about 120–130°F and close the corrugated fiberglass again. Next, run a wire from the remaining lug on the interior thermostat to the positive pole on the blower, and then connect the neutral post on the fan to a neutral "leg" in your house's wiring.

You can flip the circuit breaker on again and set the interior thermostat a few degrees higher than your main furnace control. As long as there's heat in the collector, your traditional heating system will be taking a moneysaving rest.

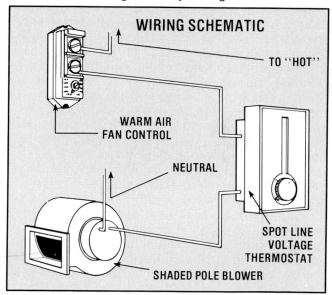

WIRING SCHEMATIC

TO "HOT"

WARM AIR FAN CONTROL

NEUTRAL

SPOT LINE VOLTAGE THERMOSTAT

SHADED POLE BLOWER

STAR POWER

The Tin Fin Press— An Indispensable Solar "Tool"

Unless you like cold showers or relish the thought of moving to Iceland or some other location where hot water flows free for the taking from springs, you'll more than likely find that the expense of providing hot water for yourself and your family is your greatest single heating-related cost next to what is spent keeping your home warm. However, this cost can be whittled down to nearly nothing with the installation and use of one of the many ingeniously designed homebuilt solar water heaters that can be fabricated by the average homeowner.

Most of the designs for sun-powered heaters include pipes through which the water flows, and these tubes are either attached to a metal plate or equipped with fins to absorb the sun's warmth and transfer that heat to the water by conduction. But making fins out of copper, as is commonly done, is expensive, and often the area shared by the plate or fins and the pipe is less than optimum. However, with an easily constructed tool designed to bend small, thin sheets of metal, the heat-absorbing wings can be made from free-for-the-gathering beverage or food cans. This tool (which can be used for copper sheets, too) helps establish the maximum heat-transferring surface between the water-carrying tubes and the warmth-gathering fins.

Conventionally, people either soldered their copper water piping directly onto a flat collector sheet, made a V-notch "trough" with angle iron for the pipework, or hired a machinist to make some conduit-hugging semicircular grooves in the back plating. Unfortunately, the first two techniques produce very little contact between the two metals, so a lot of heat gathered by the flat sheets never reaches the water pipes. And,

while the last approach works fine, it's quite expensive!

Well, this fin press was invented to solve that inefficiency-versus-cost dilemma. The bottom section of this two-piece unit consists of a 7″ length of 3/8″ copper pipe (or 1/2″ wooden dowel) that's been set—to a depth of slightly less than half its diameter—into a side-walled block. The top piece is little more than another block that has been grooved to fit over a pipe placed on the base. To work the fin shaper, just position a flat, precut piece of copper or tin on the base, put the cap over it, and push down. Presto! You have one grooved fin ready to cradle a bit of collector pipe.

The fin press can be made in only a few minutes out of scrap pipe and lumber bits, and it is simplicity itself to construct. The accompanying "fin form" illustration should be pretty much self-explanatory . . . except for the following two details: [1] To make the top press's central groove, handsaw a few starting notches, dig out the basic trough with an inexpensive round rasp, and finish shaping the groove with sandpaper. [2] The handle consists of a short piece of 2 X 4 that's been nailed or screwed onto the top board. Any sharp edges of this grip are rounded and smoothed.

To turn a collection of free cans into solar collector fins, just cut both the lids and the end rims off each one with an electric—or manual, wall-fastened—can opener (hold the tins in a horizontal, rather than vertical, position when you do this). Then—using steel wool or a propane torch—burn off the metal's protective coating. Cut each can down the middle, trim the sheet to a flat 5-1/2″ X 6″ shape, and press it.

SOLAR COLLECTORS & WATER HEATERS

When you're ready to secure a fin to some pipe, just lay the shaped tin on two fireproof boards (or blocks, or bricks) placed side by side with a pipe width's space between them. Nestle the fin's curved center section into the gap, set the collector pipe into position, hold it in place with any heavy, nonflammable weight (to increase sheet-to-pipe contact), and solder the two metals together. Add more fins to the pipe by laying out the grooved sheets, weighting, and soldering until you've completed your platemaking . . . then buff off any extra flux along the length of the pipe.

When all that's done, you'll have finished a crucial (and often the most expensive) step in the fabrication of a home-designed solar hookup, and the cost for all your collector's fins will be a grand total of zero dollars and zero cents!

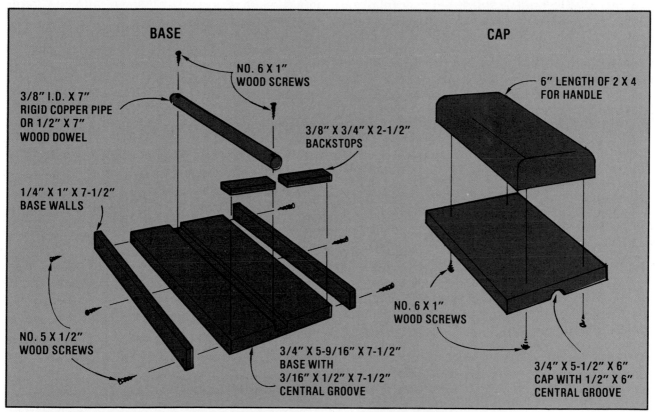

BASE

CAP

3/8" I.D. X 7" RIGID COPPER PIPE OR 1/2" X 7" WOOD DOWEL

NO. 6 X 1" WOOD SCREWS

3/8" X 3/4" X 2-1/2" BACKSTOPS

6" LENGTH OF 2 X 4 FOR HANDLE

1/4" X 1" X 7-1/2" BASE WALLS

NO. 5 X 1/2" WOOD SCREWS

3/4" X 5-9/16" X 7-1/2" BASE WITH 3/16" X 1/2" X 7-1/2" CENTRAL GROOVE

NO. 6 X 1" WOOD SCREWS

3/4" X 5-1/2" X 6" CAP WITH 1/2" X 6" CENTRAL GROOVE

STAR POWER

Turning Tin Cans Into Hot Water

Harnessing the sun's rays to heat your home's water can result in a considerable reduction in your gas or electric bill, but building a solar water heater yourself—rather than buying a ready-made unit—can mean even greater savings, especially if you construct as much of the heater as you can of free and secondhand materials. Rather than relying on expensive copper plates to snare the sun's warmth, you can fashion the heat-absorbing fins from discarded tin cans, shaping them with a homemade fin press (see page 34) to fit snugly against the collector's water-carrying pipes.

While the ten-square-foot unit you see here isn't capable of filling all of a household's hot water needs, it can put a big dent in the gas or electricity bills you've been paying for that warm water, and an enterprising fabricator could easily scale up the dimensions of this heater to increase its output.

The materials needed to build the collector are illustrated: plywood, fiberglass glazing, rippleboard, flashing, insulation, copper tubing, some 1 X 4 lumber, a collection of steel cans, soldering flux, solder, wood screws, and silicone sealant.

It takes about 56 cans—cut and trimmed to a 4″ X 5-1/2″ size—to construct a collector like this. Each of the future heat-grabbing plates should be set into the fin press so that the tube-fitting crease will be formed perpendicular to the metal's 4″ sides. Then, when the fin is soldered to a 3/8″ water-carrying tube, it will extend 2-1/2 inches on each side. By placing the four pipes five inches apart and allowing an inch of open space on each end for manifold connections, you should be able to fit 14 shaped cans on each of the collector's tubes.

As you fasten the cut containers to the 3/8″ copper tubes, be certain to sweat enough solder into each joint to insure that there will be adequate contact area to provide heat transfer from the plates to the tubes. Once all the cans are in place, bore the 7/16″ holes in the manifold pipes to accept the 3/8″ copper tubing, solder the various joints, and cap the unused ends of the manifold tubes. To position the 3/8″ tubes in the manifolds, slide a section of dowel into the large pipe, and then butt the inserted section against it.

Once all of the metal parts have been joined, the remainder of the assembly consists only of building a box around the "guts" of the collector and painting everything black. Using scrap 1 X 4 to frame the tubes, with 1″ holes drilled in the side pieces for the manifold's entry and exit, connect the six boards with countersunk wood screws. To seal the underside of the unit, simply slip three inches of fiberglass batt beneath the

The two pipes and their fittings that emerge from the collector hardly hint at the amount of soldering that took place inside the unit while it was being built.

piping, and then close in the box's bottom with a scrap of 1/4″ plywood. (Of course, the method of insulation can—and should—vary according to what materials are available. For instance, rigid polyisocyanurate insulation might actually be *more* efficient than fiberglass.)

For glazing, use the popular corrugated reinforced fiberglass which happens to be available in sizes convenient for this project. Use a liberal dose of silicone sealant, and seal the covering with ripple board on the ends and aluminum flashing on the sides.

In order to keep the cost of the project as low as possible, circulate water through the collector by natural convection. The thermisiphon approach also allows the liquid to remain inside the tubes long enough to become thoroughly warmed. However, anyone who wants to scale up the unit—by going to a full ten-foot-long section of the fiberglass glazing, for example—would probably want the collector to be served by a pump, automated controls, drain-down mechanisms, and other conveniences. The device's potential is limited only by the imagination.

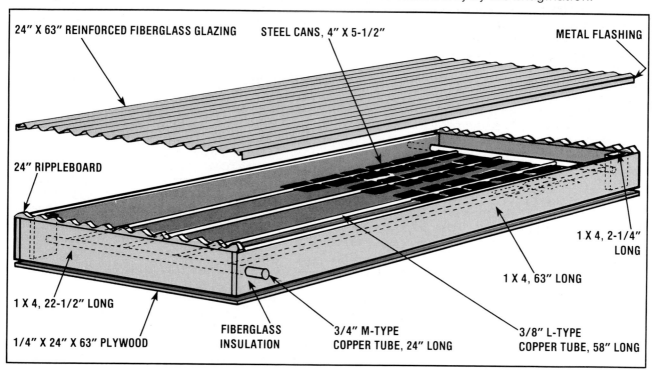

24″ X 63″ REINFORCED FIBERGLASS GLAZING   STEEL CANS, 4″ X 5-1/2″   METAL FLASHING

24″ RIPPLEBOARD

1 X 4, 2-1/4″ LONG

1 X 4, 63″ LONG

1 X 4, 22-1/2″ LONG

1/4″ X 24″ X 63″ PLYWOOD

FIBERGLASS INSULATION

3/4″ M-TYPE COPPER TUBE, 24″ LONG

3/8″ L-TYPE COPPER TUBE, 58″ LONG

STAR POWER

Turn That Old Icebox Into a "Hot Box"

Cast off after they can no longer serve to keep food and beverages cool, refrigerators all too often wind up in roadside dumps and other sites around the countryside where they become hazardous eyesores. However, with a little work and a modest investment in new materials, these throwaway iceboxes can be revived to serve as energy-saving, solar-powered water heaters, performing a task opposite the one for which they were originally designed.

Building such an appliance that incorporates a discarded refrigerator is not complicated and, in addition to the junked cooler, requires only an easily fabricated solar collector and a storage tank (the latter is put inside the used fridge). The entire unit is placed outside in the sun and generates a constant supply of virtually free hot water.

While a water heater's solar collector can vary in size and design from the one described here, it basically consists of an insulated wooden box, an absorber plate (some black-painted copper tubing on a black-painted sheet of metal), and several layers of glazing.

The frame of the collector is easy to make: Just nail together a 5″ X 2′ X 4′ box out of scrap lumber and wood paneling (the exact dimensions aren't that important).

To make the collector's absorber plate, first salvage about 15 feet of copper tubing and, after cleaning it up with sandpaper, cut the pipe into the lengths shown on the materials list. Then solder elbows to the ends of the 6″ pieces, making sure the joints face the same direction. As you put each elbow/pipe/elbow assembly on the ground, check to be certain that both ells touch the floor.

Next, solder the 18″ lengths of pipe to the elbows of the five 6″ pieces to create a zigzag pattern of pipes. This should be done on a flat surface to insure that the finished assembly is properly aligned. When this is done, solder 24″ lengths of tubing to the last two open elbows to produce the assembly shown in Fig. 1.

Now, use tin snips to cut a piece of thin galvanized sheet metal to fit the inside of the collector box. (You should also smooth the metal's ragged edges with sandpaper and round off the piece's corners to avoid accidental cuts.) Then solder the zigzag pipe assembly to the sheet of metal, punch holes in the metal along both sides of the pipe (throughout the pipe's length), and lace the assembly firmly to the sheet metal with steel wire to increase the tubing's contact with the metal. (See Figs. 2-A and 2-B.) Finally, spray-paint the entire unit flat black.

At this point, it's time to mount the absorber in the box. First, drill holes in the side of the box for the absorber's "inflow" and "outflow" pipes. Then [1] lay newspaper, polystyrene packing beads, more newspaper, and a layer of corrugated cardboard down inside the box, [2] place the absorber on top of the cardboard, placing the inflow and outflow pipes through their respective holes, [3] staple a sheet of clear plastic inside the box 1/2 inch above the absorber plate, [4] nail some small strips of wood inside the container (Fig. 3), [5] lay a cut-to-fit pane of glass on top of the wood strips, and [6] very carefully nail more strips of wood around the edges of the glass to hold it in place. (See Fig. 3.) Finally, glue polystyrene sheets to the outside of the collector and staple a large sheet of plastic over the entire assembly to keep out the winter winds.

SOLAR COLLECTORS & WATER HEATERS

An inexpensive five-gallon gasoline can makes a good storage tank for the water heater, though any vessel that'll fit inside a gutted refrigerator ought to work.

Modify the can as follows: First—after sanding away all the paint and dirt on the side of the container opposite the seam—punch four holes in the can's wall. Next, enlarge the openings to accept 1/2"-diameter copper pipe with a snug fit. Then solder 8"-long pieces of tubing to the openings so that the pipe extends 5 or 6 inches from the can. (See Fig. 4.)

And now, work on the icebox itself begins. The first order of business is to remove all accessories including the motor, the compressor, the heat exchange coils, and the freezer compartment (save the shelves). Do this outdoors and try —if at all possible—not to rupture any pipes. The refrigerant that may come spewing out of any connections you break is harmful to the earth's upper atmosphere, and it's cold enough to "burn" an arm, a neck, or a face. Work slowly, use common sense, and do the job away from children and other spectators.

Many old refrigerators have a removable panel behind the freezer area. If so, it's no problem to remove this cover, thereby creating a small port through which to insert the water tank's inflow and outflow pipes. (Note: It's important that the storage tank be located as high as possible inside the empty icebox so a good convective current will take place between tank and collector.)

For a fridge that doesn't have a removable rear panel, punch several closely spaced holes in the back of the appliance, then insert the tips of a pair of tin snips in one of the holes and cut a 3" X 12" slot (that takes in all the holes). Next, fold the cut edges under with a pair of pliers and reach in to remove the insulation from the cutaway portion of the wall. On the other side of the wall, cut away a second 3" X 12" opening (again, turn the ragged edges under). The water tank's pipes should fit through this slot. (See Fig. 5.)

Now, find a block of wood that's a little larger than the slot in the back of the refrigerator. Drill four holes in the wood corresponding to the four pipes in the water storage can, then drill holes for bolts around the periphery of both the block and the fridge's 3" X 12" slot. Bolt the block to the opening in the back of the icebox.

It's time now to insulate the storage tank. First, wrap the tank with aluminum foil (shiny side in) and then wrap its sides and bottom with whatever insulation is available. Several layers of scrap plastic and two days' worth of newspapers bound in place with steel wire will do the job. Be sure—before going any further— that your storage tank has a vent hole in its top. Make one, if necessary.

You might—at this point—also want to insulate the refrigerator's door by gluing an inch of plastic foam sheeting to the inside of it.

All that's left is to mount the water storage tank inside the refrigerator. The tank can rest on one of the existing shelves, or you can rig a makeshift support. Finally, neatly fill the interior of the icebox with old newspapers in such a way that the door can be opened later without making a huge mess. (See Fig. 6.)

Assuming that the water tank is to be filled manually, you'll want to put a 90° elbow and a short piece of tubing on the tank's topmost—or number one—pipe, as shown in Fig. 4. (Note that the short piece of tubing must not extend any

STAR POWER

Turn That Old Icebox
Into a "Hot Box" (continued)

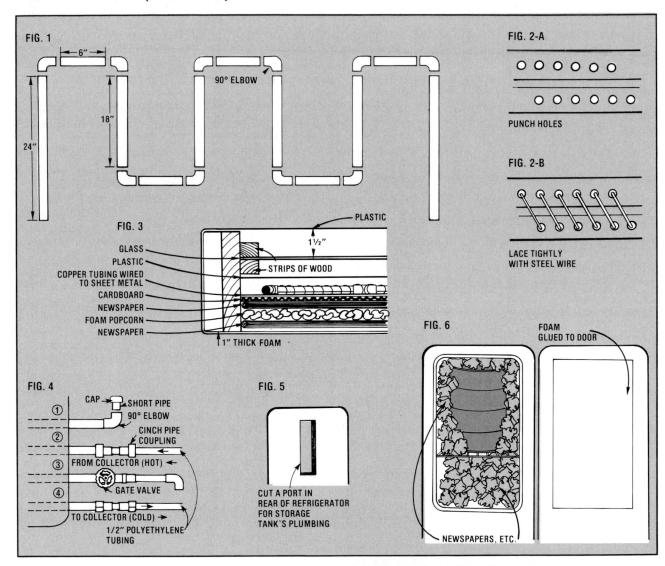

FIG. 1

6"

18"

24"

90° ELBOW

FIG. 2-A

PUNCH HOLES

FIG. 2-B

LACE TIGHTLY
WITH STEEL WIRE

FIG. 3

PLASTIC
1½"
GLASS
PLASTIC
COPPER TUBING WIRED
TO SHEET METAL
CARDBOARD
NEWSPAPER
FOAM POPCORN
NEWSPAPER
STRIPS OF WOOD
1" THICK FOAM

FIG. 4

CAP
SHORT PIPE
90° ELBOW
①
CINCH PIPE
COUPLING
②
③ FROM COLLECTOR (HOT)
GATE VALVE
④
TO COLLECTOR (COLD)
1/2" POLYETHYLENE
TUBING

FIG. 5

CUT A PORT IN
REAR OF REFRIGERATOR
FOR STORAGE
TANK'S PLUMBING

FIG. 6

FOAM
GLUED TO DOOR

NEWSPAPERS, ETC.

LIST OF MATERIALS

Plastic foam or fiberglass insulation
Newspapers
Polyethylene plastic
Polyethylene tubing
One 5-gallon container (new or well cleaned)
Flat-black paint (spray and brush-on)

SOLAR COLLECTORS & WATER HEATERS

higher than the vessel's vent opening, or the tank will overflow when it's filled.) Keep a 1/2" end cap on the pipe between fillings to insure that the system's water will remain unpolluted. (Run a small length of chain or wire from the cap to the wooden block on the back of the fridge, and you'll never have to worry about the cap becoming lost, stolen, or contaminated.)

The second and fourth pipes from the top of the storage tank will lead to the collector. To these pipes' ends, attach plastic connectors of the type called cinch pipe. (This type of connector allows a person to fasten and unfasten polyethylene tubing quickly and easily and is a big plus when it comes to moving the solar heater's components around.) If you plan to move your heater often, you'll probably benefit from using cinch pipe connectors. (You might want to put cutoff valves on each of the storage tank's three lowermost pipes, too, to avoid getting wet when you disconnect the hoses from the tank.) Otherwise, plumb the tank with copper pipes and fittings and count on leaving the system at a permanent location.

After putting cinch pipe fittings on the tank's number two and four "spouts", [1] affix the same kind of fittings to the collector's inflow and outflow pipes, [2] put a piece of polyethylene tubing between the number two pipe and the topmost collector fitting, and [3] run a piece of plastic tubing from the tank's number four pipe to the collector's lowermost cinch pipe fitting.

The number three spout is to be the hot water faucet. This pipe should be fitted with a gate valve followed by a 3" or 4" segment of pipe and a down-facing 90° elbow. Be sure you mount the valve so that the little arrow on the casting faces in the same direction as the flow, which is *away* from the storage tank.

Now, to get the water heater set up and working, just [1] place the collector where it'll be in constant sunlight, [2] face the panel due south, [3] angle it to receive the maximum amount of sunlight (at an angle from the ground equal to your latitude plus 15°), and [4] fill the system with water. Several hours later (assuming the sun's out), the water will be hot.

Just *how* warm the water will be depends on the size of your collector, the outside air temperature, and other factors. But in the case of one experimental unit, records show the water temperatures soared into the 140° to 160° F range on sunny summer days, and dropped only to 70° to 90° F on the coldest sunny winter days. And that was with a little old eight-square-foot collector!

LIST OF MATERIALS (continued)

12 linear feet of 1 X 5 pine (or equivalent):
two 2'-long pieces
two 4'-long pieces
One 2' X 4' piece of 1/4" paneling, hardboard, or plywood
One piece of rolled, galvanized sheet metal, approximately 2' X 4'
10 copper ells for absorber plate
3 to 5 copper ells for plumbing

15 feet of 1/2" copper tubing:
two 24"-long pieces
four 18"-long pieces
five 6"-long pieces
four 8"-long pieces

20' to 30' of steel wire
4 cinch-pipe tubing connectors
One pane of glass, approximately 2' X 4'
Nails and/or wood screws

STAR POWER

Thermosiphoning— The Solar Hot Water "Pump"

In our technologically advanced country, it's all too easy to get in the habit of assuming that complex solutions to problems are more thorough and, therefore, somehow superior to less complicated approaches. Take the usual attempts to capture energy from the sun, for example: A typical commercially made solar domestic hot water system has—as a minimum—a pump, several collectors, a maze of plumbing, numerous temperature probes, and at least one differential thermostat.

But are all those components really necessary? No! It's entirely possible to build an effective solar water heater that doesn't require a single watt of electricity to circulate the liquid.

Liquid in such a setup circulates through the courtesy of an everyday physical phenomenon: In a body or a container of liquid, the warmer fluid rises to the top, and cooler fluid falls to the bottom. This occurs because when matter is heated, its density is reduced, and the influence of the force of gravity on a given volume of a

When mounted on the south side of a building and connected to a storage tank, the collector operates quietly and continuously when the sun is shining. Cold water enters the bottom of the solar water heater through the tube at the right and heated fluid exits at the top left of the array of collectors.

SOLAR COLLECTORS & WATER HEATERS

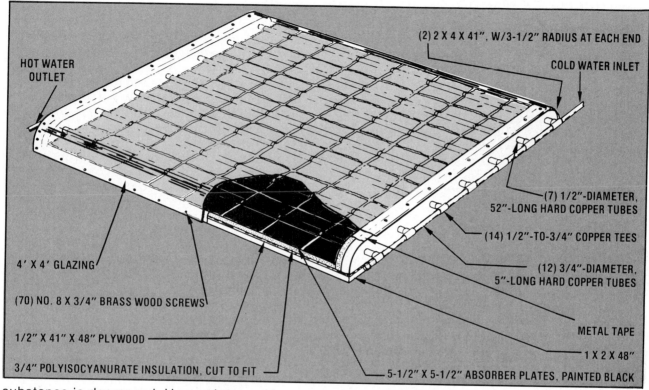

HOT WATER OUTLET

(2) 2 X 4 X 41", W/3-1/2" RADIUS AT EACH END

COLD WATER INLET

4' X 4' GLAZING

(70) NO. 8 X 3/4" BRASS WOOD SCREWS

1/2" X 41" X 48" PLYWOOD

3/4" POLYISOCYANURATE INSULATION, CUT TO FIT

(7) 1/2"-DIAMETER, 52"-LONG HARD COPPER TUBES

(14) 1/2"-TO-3/4" COPPER TEES

(12) 3/4"-DIAMETER, 5"-LONG HARD COPPER TUBES

METAL TAPE

1 X 2 X 48"

5-1/2" X 5-1/2" ABSORBER PLATES, PAINTED BLACK

substance is decreased. Hence, in a passive solar water heating system where the storage tank is situated above the heat collectors, a circulation known as a thermosiphon occurs as the warm water rises from the collector to the tank, and cooler water from the bottom of the tank flows back down to the suncatcher where it is heated and rises again.

Because the "pump" of a thermosiphon setup is powered by solar energy, the intensity of the sun's radiation regulates the rate of flow. In fact, when the sun stops shining on the collectors, the system is effectively shut off. However, though the concept of thermosiphoning is surprisingly simple, there are a few basic rules to follow when designing such a collector.

First, the convective loop will work properly only if the bottom of the storage tank is at least a foot and a half above the top of the collector. This head is necessary to build pressure in the system and to help prevent a reverse flow from occurring at night.

STAR POWER

Thermosiphoning—
The Solar Hot Water "Pump" (continued)

Furthermore, the connections to the storage tank need to be properly located in a convective loop, solar hot water setup. The cold line to the collectors should exit at the bottom of the tank, and the hot return must enter near the top. Also, the vertical distance between the cold line's exit from the storage tank and its entrance to the collector must be less than the comparable distance from the exit of the hot line from the collector and its entrance to the tank.

Because thermosiphon systems rely on relatively weak convective forces to provide their circulation, they need to be built with generously sized tubing to reduce friction against the pipe walls. Depending on the distance between the collector and the storage tank, the pipes should be at least 3/4″ in diameter. Also, all pipes must be thoroughly and equally insulated, and the feed and return lines should be pitched evenly from the collector to the tank without any significant dips that could catch air bubbles.

In collectors similar to the ones shown, either the units should have their headers built in at a slight angle, or the entire collector should be tilted slightly upward toward the side of the hot water outlet.

These collectors for the thermosiphoning system suit the convective arrangement better than many in-line units. The two sun grabbers also make efficient use of conventionally sized materials because a pair of collectors can be made from one sheet each of plywood and insulation.

The boxes are 41″ wide and 48″ high and are identical in all respects except for their absorber plates. One is made by using stapled aluminum printing-plate absorbers and the second with copper sheets soldered to the tubes.

Each collector frame is made from a half-sheet of 1/2″ CDX plywood and a half-sheet of 3/4″ polyisocyanurate insulation, a 2 X 4 X 8′ cut in two—with the ends of each piece trimmed to a 3-1/2″ radius and with 1/2″ holes drilled on 5-7/8″ centers down its length—a 4-foot-square piece of fiberglass-reinforced plastic glazing, a 1 X 2 X 8′ cut in half to serve as side pieces to which the glazing can be screwed, foil tape, and 70 No. 8 X 3/4″ brass wood screws.

Plumb the units with seven 52″ lengths of 1/2″ copper tubing apiece, and connect those tubes to 52″-long 3/4″ headers. To do this, cut the 3/4″ pipes into approximately 5″ sections and sweat solder the array together, using 1/2″-to-3/4″ copper tees.

As was mentioned, the two collectors shown here have different absorber plates. One is fitted with 5-1/2″ X 5-1/2″ sheets of 0.010″-thick copper (about 4 square feet altogether), while the other has twice as many like-sized sections of 0.010″-thick aluminum (or a total of about 8 square feet, though the areas exposed to the sun are about the same). In both systems the plates have grooves pressed into them with a homemade fin press (see page 34), but the copper is sweated to the 1/2″ tubes while the aluminum sheets are sandwiched around them and then simply stapled together.

Preliminary results showed that the simpler, less expensive aluminum sandwich system performed almost as well as the more costly, more difficult-to-construct copper fin setup. In fact, the unit equipped with printing plates produces a temperature rise that's within a degree of the increase accomplished by the more conventional collector.

SOLAR COLLECTORS & WATER HEATERS

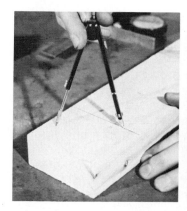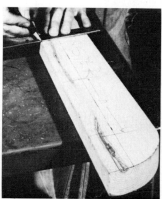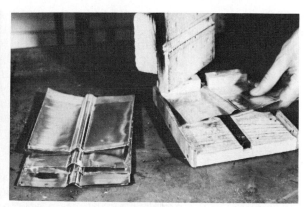

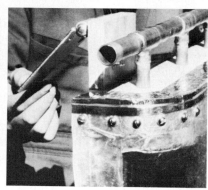

Some of the steps involved in fabricating the collectors shown here include setting a compass to the width of a 2 X 4 and scribing an arc from a corner of the board to the edge, trimming along that line and then marking off centers 5-7/8 inches apart for the holes the pipes will go through, pressing a groove in the collector plates, and fastening the plates to the tubes (aluminum plates are stapled around the tubes in pairs while copper plates are soldered to the pipes). The plates and tubes together form the sun-catching surface which is then encased in the collector housing. The completed unit can be mounted to a wall or overhang by means of hinges mounted to the heater's back.

STAR POWER

A Simple
Solar Shower

There are numerous specialized tasks that the sun is just burning to undertake for people who—with a little ingenuity—invent ways to harness even a tiny portion of the virtually limitless solar energy that falls upon the face of the earth. Warm water for bathing is considered almost a necessity these days, and with a solar-heated shower, that hot water can be provided without the "benefit" of the expensive assistance offered by the gas company or the local electric utility.

Designed for use at a summer cabin on an island off the coast of Maine, the model here consists of little more than a roof-mounted storage tank/collector and the hose and pipe fittings needed to bring liquid to the tank and to make the heated water available to a shower stall in the living area below. The collector assembly can be put in place in the spring and removed and stored in the fall.

Almost everything necessary for fabricating a sun-heated shower can probably be found at your friendly neighborhood dump or at a good secondhand store. The storage tank should be a galvanized unit and should be painted black. The dimensions can vary so long as the capacity is about 15 gallons. To provide a means of mounting the tank on the roof, solder two pieces of perforated galvanized pipe-hanging strapping, two to three feet long each, to the bottom of the tank. (All soldering should be done in a well-ventilated area, since soldering flux and galvanized materials give off toxic fumes when heated.) Bend the straps so they will fit over the ridge of the dwelling, and so that one of the holes in each hanger can be hooked over a headless nail driven into the roof on the side opposite the container.

Next, punch two holes—each about 3/4" in diameter—in the side of the tank that will, when it's in place, be nearest the eave. Position one of these punctures near the bottom of the tank and the other near the top. With that done, solder 3/4" galvanized pipe couplings over the holes, and thread a 6"-long 3/4" galvanized pipe nipple into each one. Then thread a 3/4" 90° galvanized elbow onto the end of each nipple and tighten these until the elbows are facing down. Now, screw a 3/4" X 3/4" CPVC male-pipe-thread–to–hose-barb adapter into each of the elbows. On the adapter coming from the bottom end of the tank, install a 4" length of 3/4" flexible hose, securing it with a hose clamp. Go on to insert a 3/4" CPVC barb "T" into the other end of this pipe with the stem of the "T" facing the uphill end of the tank, and tighten this joint with another hose clamp. Then connect, to the stem of the "T", a length of 3/4" hose that is long enough to reach a short distance above the uppermost point of the tank to act as an overflow and pressure-release valve.

Moving to the adapter at the upper end of the heater, attach a length of hose long enough to reach to the shower stall in the house below, and fit an old-fashioned "collander" shower head to the "business end" of the system. Fasten both the joint at the tank and the one at the shower head with hose clamps. Connect a second long length of hose from the vacant arm of the "T", run the length of tubing to a valve-fitted cold-water supply line in or beside the shower stall, and attach it with an adapter. (Hose clamps are needed at the joints at each end of this pipe, also.)

To complete the rooftop tank/collector, place a rectangular frame built of scrap 1 X 6 lumber

SOLAR "APPLIANCES"

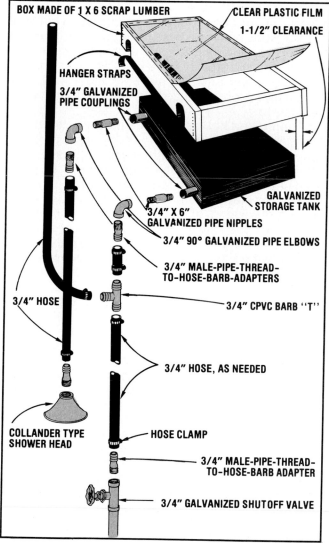

BOX MADE OF 1 X 6 SCRAP LUMBER

CLEAR PLASTIC FILM

1-1/2" CLEARANCE

HANGER STRAPS

3/4" GALVANIZED PIPE COUPLINGS

GALVANIZED STORAGE TANK

3/4" X 6" GALVANIZED PIPE NIPPLES

3/4" 90° GALVANIZED PIPE ELBOWS

3/4" MALE-PIPE-THREAD-TO-HOSE-BARB-ADAPTERS

3/4" CPVC BARB "T"

3/4" HOSE

3/4" HOSE, AS NEEDED

COLLANDER TYPE SHOWER HEAD

HOSE CLAMP

3/4" MALE-PIPE-THREAD-TO-HOSE-BARB ADAPTER

3/4" GALVANIZED SHUTOFF VALVE

(be sure to include cutouts for the pipes on one side) topped with plastic glazing.

To fill the tank, simply turn on the cold water valve, which is the rig's only control. When water begins to flow out of the shower head, the storage unit is full. Once the water has been warmed by the sun, all a would-be bather has to do is step into the stall and gently turn on the valve. The cold water flows into the bottom of the already full reservoir, forcing the heated water into the pipe at the top of the tank and down to the shower head. Because the tank is about ten feet above the bather's head, the water will continue to run for a short time after the valve is shut off, usually just long enough to perform the rinse cycle. This height offers enough gravity pressure for comfortable bathing and rinsing, but not so much pressure that water is needlessly wasted.

At the cottage in Maine, the heater's water temperature averages around 85°F on overcast days and reaches as high as 110°F on sunny days. In very warm weather, the water becomes too hot for midafternoon bathing, but this isn't a major drawback for those people who don't mind scheduling showers at noon before the water becomes intolerably warm or in the evening after it's had a chance to cool off somewhat. The reservoir holds enough water for three generous or four skimpy showers . . . even if they're taken one following another.

The entire appliance, even if all the pipe fittings and galvanized strappings are purchased new, should cost very little to construct. Finagling to get as many free components as possible will prove even more economical. Either way, several summers of fuel-free hot water will justify the amount you spend.

STAR POWER

Taking It With You— A Portable Solar Water Heater

Most of us have—at one time or another—longed for the luxury of a hot shower while on the road or trail. Well, there's no need to yearn any longer. Here's a portable solar water heater that costs very little and can be folded and carried in a backpack, suitcase, car, or boat!

It's also quite easy to fabricate this heating unit using the following supplies: one auto inner tube, a length of garden hose, a 1/2″ plastic pipe coupler, a small clamp, a male hose connector, a hose valve or nozzle, and some quality rubber-sealing compound.

It's usually possible to scrounge a free, secondhand inner tube in good condition, but if not, buy a used one from a tire shop. Keep in mind that a 14-inch tube will hold around 19 quarts of water. For greater or lesser capacity, simply go up to a 15-inch truck tube or down to a 13-inch compact-car size.

Once the tube proves to be sound (inflate it to find out if it has any leaks), use a sharp pair of scissors to cut a 1/4″ hole about two inches in from the outside edge. Now, wet your finger with water and clean around the inside of the hole.

Next, take the 1/2″ pipe coupler, and leaving the small ridge, cut off its lower half. Then apply a good amount of rubber-sealing compound on that ridged end, and push it into the 1/4-inch hole until the ridge is inside the tube. (This task may require a little effort, but it can be made easier by leaving the tube in the sun for a while to soften up the rubber.)

With the lower part of the coupler in the hole, pull the ridge up snug against the tube wall and let the cement dry for the maximum time stated on the product's label, because if it doesn't dry properly, the seal will be worthless.

When that is done, clean around the "new" stem and roughen that area just as you would the space surrounding a hole in an inner tube that's to be patched. Put some rubber-sealer on your finger and coat the outside area. Three thin coats will dry faster and provide a stronger seal than one large glob. Again, allow the cement to set completely.

Then take about two feet of old garden hose, connect it to the new stem (if there's a problem slipping the hose on, let it warm in the sun for a while to soften), and put a clamp around the hose and stem to hold them together.

Finally, attach a standard nozzle or hose valve to the other end of the hose. Try to use an old piece of hose that still has the male connector on it to which the water control valve can be fastened. Otherwise, purchase an inexpensive garden hose repair kit, which will contain the fitting you need.

Using this portable water heater is even easier than making it. A funnel will speed filling the tube through its hose, and a bucket will help when the water supply is a lake or stream.

Once the tube is full, place it in the sunshine where the water's temperature will be raised about 10° every hour, leveling off at about 115°F. For a faster heat gain, place the "tank" on the hot roof of a car or truck that's parked in the sun. It's a simple matter to draw water from the heater in such an elevated position, too.

When you're ready to move on, merely empty the inner tube and roll it up for handy storage. But don't forget about it when you get home: In addition to camping and traveling uses, the rubber doughnut can be a convenient water source for outdoor hand-washing.

A pipe coupler, cut off just below its middle ridge, is sealed into a hole in a used inner tube. The tube filled with water is warmed by the sun and, with a hose attached, becomes a handy source of hot water while camping or traveling.

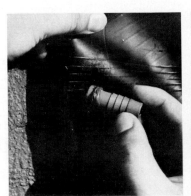

SOLAR APPLIANCES

The Cardboard Box Food Dryer

Many people in both government and industry recognize the value of harnessing solar energy and putting it to work in place of, or in addition to, power derived from nonrenewable fossil fuels. But when they "talk solar", they are invariably concerned with multimillion dollar projects, brain-trust developed technology, and decade-long implementation schedules.

All this may be necessary on a national or even global scale, but individuals can utilize the sun's energy, too, with easily accomplished projects such as the solar food dryer shown here, which anyone can put together very inexpensively and in just a little time.

To build this sun-powered unit, choose a long, shallow cardboard box and cut a few holes in each of the narrow ends. Next, paint the inside of the container black or line it with black plastic sheeting, whichever is easier. Cover the box with clear plastic, and the heat catching portion of the dryer is complete.

Now, take another cardboard box and make several holes in one side. Using tape to hold the material taut, cover the open top with a cloth screen on which the food to be dehydrated will be placed. Set the drying box on a table or stand, and lean the solar box against it at the most effective angle to catch the sun's rays. Then use some scrap cardboard and masking tape to form an air duct from the angled space between the two units.

By really splurging on this device, a person might be able to spend as much as a dollar. Granted, it's not very permanent, and it's certainly not waterproof ... but who would want to leave drying comestibles out in the rain or snow anyway?

DRYING BOX

CLOTH SCREEN

BLACK INSIDE

CLEAR PLASTIC SHEET

AIR FLOW

SOLAR BOX

STAR POWER

A Stellar Broiler for *Real* Outdoor Cookery

There seems to be no end to all the new single-function appliances on the market. There are hot dog cookers, bagel makers, bacon fryers, and on and on. Offering a wide selection of costly gadgets may be brilliant marketing, but it's hardly economical for a person to buy them all, especially since most of them can be homemade.

Here's one item that will add a little extra fun and excitement to those outdoor picnics and cookouts that your family enjoys. It's a solar hot dog or shish kebab broiler built of wood, a few pieces of hardware, and some flexible mirrored or reflective sheeting. When aimed at the sun, the unit catches the solar radiation that strikes its concave face and focuses it into a concentrated line of heat that falls on the broiler's skewer. The whole shebang can be put together inexpensively and once completed, costs nothing to operate.

The trick to making the cooker work properly lies in designing the collector's curved back so that it focuses the sun's rays into one narrow and intense band of heat. But that's not as difficult as it may seem, and the parabolic curve can be laid out on one of the 2 X 8's with a nail and a framing square as shown in the accompanying sketch.

Using the nail as a pivot, swing and slide the framing square so that its corner moves precisely along the edge of the 2 X 8. Stop the square often to draw a series of overlapping lines as shown, which will describe the necessary curve.

Once it's drawn, cut the curve with a band saw, saber saw, or coping saw, and smooth it with either sandpaper or a rasp. Then use what's left of the first 2 X 8 as a pattern to mark and cut a second. These two rounded pieces of wood will be the cooker's sides.

Cut a 1/4"-square notch (centered right on your parabolic curve's pivot point) out of the straight edge of each of the sides. The skewer will fit into these notches when the cooker is finished.

Then cut a bigger notch 1-1/4" deep and 1-1/2" long into one of the sides as shown. Drill a 1/4" hole one inch deep into the back of this indentation taking care to make the hole perpendicular to the face of the piece of wood. Finally, put a little glue on one end of a 1/4" X 2" wooden dowel and push or tap this "aiming" peg as far into the hole as it will go.

Next, glue and nail the 1/8" X 26" X 32" piece of paneling—trimmed to fit properly—to the

SOLAR APPLIANCES

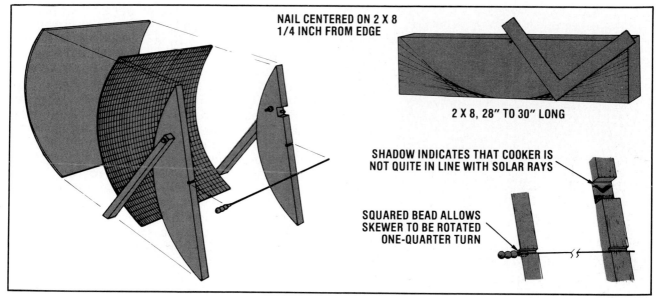

NAIL CENTERED ON 2 X 8
1/4 INCH FROM EDGE

2 X 8, 28" TO 30" LONG

SHADOW INDICATES THAT COOKER IS
NOT QUITE IN LINE WITH SOLAR RAYS

SQUARED BEAD ALLOWS
SKEWER TO BE ROTATED
ONE-QUARTER TURN

curved edges of the two wooden sides. When the back is in place, trim the flexible mirrored sheeting so it fully lines the paneling, and glue it in place.

Make legs for this little sun scoop by drilling the necessary holes and bolting the two 3/4" X 1" X 20" pieces of pine to the cooker's sides as illustrated. The skewer is fashioned by gluing one small and three larger wooden beads onto a length of 1/8" (or larger) stainless steel welding rod. File four flat spots on the small bead once the glue is dry. When the squared bead is placed in the skewer-holding notch, broiling goodies can be rotated as they cook.

Add a coat of nontoxic paint to all the sun scoop's wooden parts, and it's ready to start cooking. Spear three or four wieners or any de-

sired shish kebab fixings on the stainless steel rod, set the skewer into its notches, and point your broiler at the sun. It'll be aimed correctly when the little round peg in the square notch casts no shadow at all. Reposition the cooker as necessary as the sun moves across the sky.

On a good day, hot dogs will be steamed tender and juicy all the way through in only four to five minutes, though a shish kebab will take a little longer. Don't worry about juice that drips onto the mirrors. It can be easily wiped off when you're finished. Or you can collect all those delicious drippings (and also increase the broiler's effectiveness on marginal days) by painting some aluminum foil flat black on one side and tightly wrapping it, painted side out, around the franks or shish kebab.

LIST OF MATERIALS

Two 2 X 8's 28" to 30" long
One piece of 1/8" X 26" X 32" paneling
Two pieces of 3/4" X 1" X 20" pine
One 1/8"- or 3/16"-diameter stainless steel welding rod
Four (three large, one small) wooden beads to fit the rod
One 1/4" X 2" wooden dowel
Two 1/4" X 3" carriage bolts, with nuts and flat washers
Glue and paint as needed
Flexible mirrored or reflective sheeting to line paneling

STAR POWER

A "Celestial Substation"
Drawing Electricity From the Sun

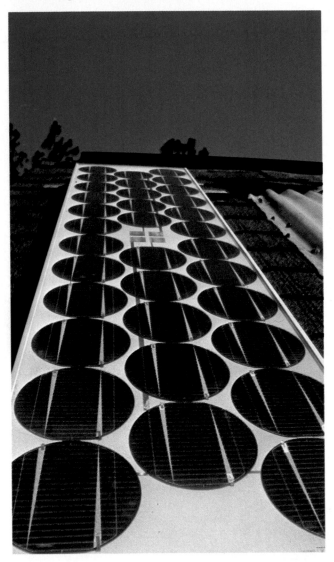

In addition to the many ways in which the sun's rays can be collected and put directly to work warming homes in the winter or heating water for bathing and washing all year round, solar energy can be used indirectly when converted to electricity, which can then be put to work producing light, powering appliances, and performing a myriad of other tasks.

Many people are inclined to be skeptical at the mention of *electricity* from sunlight: They assume that photovoltaic setups are impractical, expensive, and something only for the future. Not so, because for many people, practical solar electric power is here today at affordable prices.

Changing sunlight into electricity—the process called photovoltaic conversion—was pioneered by Bell Laboratories in the mid-fifties. And the silicon solar cells that were first developed for the space program are still the workhorses of that industry.

These cells are sliced from a cylinder of ultra-pure silicon crystal, which is nothing more than highly refined sand. Every wafer is then chemically treated and processed to form a semiconductor junction (the technique is similar to that used in the fabrication of common transistors). It's within this thin semiconductor junction that electricity is generated.

And just how is the power produced? Well, basically, light strikes the junction, liberating electrons, though the action involves a mechanism that can be fully explained only by an excursion into quantum physics. The freed electrons are then collected on a conductive grid placed over the face of the cell. When a wire is connected from the front grid to the back of the cell, an electric current flows.

A photovoltaic panel is a series of silicon cells that transform the sun's energy into electricity. A power inverter converts the DC voltage produced by a photovoltaic system into generally more usable AC voltage.

ELECTRICITY FROM THE SUN

Each cell generates about 0.5 volt of direct current electricity, while the amount of current (amperage) depends upon the size of the cell and upon the number of free electrons, which is proportional to light intensity.

Since each unit is capable of producing only about 0.5 volt, the cells must be connected in a series circuit in order to increase the voltage to a useful level. (The procedure is similar to stacking flashlight batteries.) In theory, then, 24 cells will give a total output of 12 volts. In actual practice, however, each cell's output is closer to 0.46 volt, so 26 cells are required to produce a full 12 volts. And, though amperage varies from manufacturer to manufacturer (depending on the efficiency and size of the cells), a typical 12-volt panel might produce 2 amperes.

If left *unprotected*, silicon photovoltaic cells would be susceptible to damage from moisture and airborne contaminants. So, after they're wired together, the wafers are laid face down on a sheet of safety glass. A piece of plastic (such as Mylar) is then stretched across the back of the assembly and heat-bonded. Last of all, the 3/8''-thick panel is crimped into a metal frame, both to protect the glass and to help conduct heat away from the cells. A perfect seal is then insured by applying a liberal bead of silicone sealant along the joined edges.

Currently, the market prices for panels vary from $10 to $20 or more per watt of capacity . . . that is, a 30-watt panel would cost between $300 and $600. But remember . . . that's for prime, highest quality collectors.

There are ways, fortunately, to purchase panels for *less* money. One possibility is to buy surplus equipment. Because the photovoltaic industry is expanding so rapidly, today's top seller may be replaced by an improved version at any time, and the obsolete units often sell for less than $10 per watt. Look for existing photovoltaic systems that are being updated—the U.S. Department of Energy has a few scattered throughout the country—or check directly with manufacturers to find out whether they have any obsolete panels in stock.

"Manufacturer's seconds" (panels whose performance isn't up to one or more of the maker's specifications) can also be purchased at reduced prices. The most frequently found defect in such units is the production of a lower current output than was expected, although in rare cases a defective cell may reduce panel voltage, too. You might even be able—if you can do business with a vendor who has a government contract—to acquire panels that have been rejected merely for cosmetic reasons (such as discoloration or blemishes that in no way affect the performance of the units).

If you shop prudently, you can probably find imperfect collectors for as little as $5 per watt. And, if you are able and *qualified* (or know someone else who is) to inspect them before you lay your money down, seconds may prove to be the best way to start a home photovoltaic setup. But be forewarned, the increased interest in solar electricity is rapidly drying up the surplus-and-seconds market. Therefore, bargains are getting harder and harder to come by, and you'll have to do your homework.

There is, however, still one more way to save on the cost of solar cells: quantity buying. Because manufacturing expenses drop dramatically with increased production, companies are

STAR POWER

A "Celestial Substation"
Drawing Electricity From the Sun (continued)

usually willing to give a significant discount on large orders. As a matter of fact, as much as 50% can be lopped off the sticker price when groups of homeowners buy cooperatively.

But why would someone opt to use solar electricity in the first place? Even at $10 per watt, a photovoltaic system can hardly compete with readily available utility power. (Of course, folks who are facing steep installation charges for long service-entrance lines may find that solar cells are already a bargain.) But consider for a moment: Oil prices are unstable, and more than two-thirds of the electrical generation capacity in the United States is petroleum-fueled. Photovoltaic cells, on the other hand—while subject to short-term price fluctuations—are generally becoming *less* expensive. Many experts think that the costs associated with the two systems will be equal before the turn of the century, and some even believe that the prices will balance out within five years!

The point is that *today* is a good time to begin building the groundwork for your home photovoltaic system by setting up a small powerplant that can be expanded as panels become less expensive.

A basic solar-electric system consists of nothing more than a photovoltaic collector and a load. Such arrangements are commonly employed to pump water in remote areas. By referring to Fig. 1 (on page 57), you'll notice that the wires from the solar panel connect directly to the motor. When sunlight strikes the collector, it generates electricity, which in turn powers the pump. Now, to get acceptable performance and reliability out of such a setup, it's important to be sure that the pump's motor is compatible with the panel's output. The voltage must be the same, and the collector must be capable of supplying enough current to match the pump's rated capacity. However, in order to make such comparisons, you'll have to know just how the demand and the output are related.

Photovoltaic panels are, for technical purposes, rated in volt/amps rather than watts. One volt at 1 amp equals 1 volt/amp, and consequently, 12 volts at 1 amp will work out to 12 volt/amps. Of course, the motor used in a pumping system may be rated in watts, but it should also have separate voltage and amperage ratings.

Let's say that we want to use a motor that has a listed capacity of 12 volts at 5 amps. First, keep in mind that most commercial panels are standardized at 12 volts, so the voltage will likely match. Next, assume we have a 2-amp panel, which results in a collector with a 24-volt/amp rating (12 volts times 2 amps equals 24 volt/amps). The motor requires 60 volt/amps (12 volts times 5 amps). Thus, we'll need three panels hooked in parallel, as shown in Fig. 2, to operate our pump properly. And obviously, in order to expand this system, we need only add more panels in parallel to achieve any current level necessary for a specific job.

But can a low-voltage collector be practical? Well, there are numerous devices that operate on 12 volts—auto radios and stereos, small motors, and recreational-vehicle refrigerators are just a few examples—but there are also a number of appliances that just won't work at such a limited voltage.

However, the voltage of a photovoltaic system can be increased by connecting panels in series, as shown in Fig. 3. You must remember, though, that amperage will seek the level of the weakest panel in the group. Therefore, if you hook a 1-amp and a 2-amp panel in series, the resultant output will be only 1 amp. But if two panels of 2-amp rating are hooked in series, the output will be 2 amps.

Obviously, connecting panels in series can make solar electricity much more versatile than can a simple parallel setup. There's a great deal of 24-volt, 36-volt, and 48-volt equipment available. Furthermore, if you hook nine panels in series, you'll have a 110-volt DC unit, and many common 110-volt AC appliances will operate on DC also. (Small hand tools, kitchen gadgets, heating elements, light bulbs, and radios and televisions specifically claim compatibility with both alternating and direct current.)

In addition, groups of series-connected collectors can be wired in parallel, as long as they're stacked in increments of equal total voltage. In Fig. 4, notice that three panels at a time must be added to a 36-volt system . . . no more, no less!

But what happens when the sun doesn't come out? Well, solar-electric collectors continue to produce power on cloudy days, but at about 50% of their ''normal'' rate. And, of course, the semiconductors rest each evening. Consequently, while a power storage system isn't absolutely necessary, it certainly can be useful.

There are numerous ways of storing electricity, but the lead/acid battery is the least expensive and most widely available. And, although we've referred to panel output as 12 volts up to this point for the sake of convenience, nearly all manufacturers have had the foresight to design panels that produce the 14 to 16 volts necessary to charge a 12-volt storage battery.

The batteries needed for photovoltaic systems are the deep-cycle type commonly employed to provide storage for windplants and power for electric vehicles. They're different from standard automobile batteries in that they are designed to withstand numerous discharging/charging cycles. The storage cells are available in 2-volt, 6-volt, or 12-volt units.

In the same way that panels can be grouped to produce the output required, any number of batteries may be connected in parallel as long as the voltage of each matches that of the system. And when wired in series, the batteries again need to match the panel voltage (see Fig. 5).

To prevent the deep-cycle units from overcharging, a specially designed regulator can be placed in line with the electricity-generating circuit. However, if you set up a well-balanced system—one which consumes the same amount of power that it generates—a charge regulator shouldn't be necessary.

Some electrical devices, such as powerful electric motors and color television sets, can't be operated on direct current. Fortunately, you can change photovoltaic DC power to alternating current by using an inverter. Such a device employs a pair of switching transistors to change the direction of the current 60 times per second. This form of electricity is known as 60-cycle alternating current. The AC electricity is then run through a step-up transformer to reach household voltage.

Although you can find inverters in a wide variety of power ranges, units of less than 500 watts are usually the easiest to locate and the least expensive. The smaller inverters can get by with 12-volt input current (as shown in Fig. 6), while

STAR POWER

A "Celestial Substation"
Drawing Electricity From the Sun (continued)

higher-wattage inverters generally require substantially greater input voltages to overcome internal losses caused by the larger unit's heavier circuitry. For example, a 48-volt input is common for a 2.5 kw inverter. Consequently, when mapping out an expandable photovoltaic installation, a series/parallel arrangement needs to be designed with future inverter purchases and the possible addition of other equipment in mind.

"That's all very interesting," you may be saying, "but won't it still cost me an arm and a leg to get started in photovoltaics?" The answer is that it might, except that solar electricity has one further remarkable property: It is perfectly compatible with, and can supplement, other electricity-generating systems.

Let's say, for instance, that you have a 500-watt wind charger, and the system hasn't quite been able to meet all your needs. Replacing it with a larger, more powerful generator would probably require a new tower in addition to the new windplant. In such a case, it's quite probable that solar electricity could supply the needed extra energy at a substantial saving. The panels can be made to match the wind generator's voltage by linking the solar units in series. And if your wind charger happens to be a 36-volt model, the addition of three panels in series will almost double your power!

Now that claim might sound a bit farfetched. Bear in mind, though, that windplants stall in periods of calm or in relatively low breezes— such as often occur in the middle of the day— during which time you must rely on stored power. But solar cells are at their peak at noon, so the panels can fill in for the idle wind machine. And the reverse is frequently true under low light conditions: During a storm the sun doesn't shine much, but the wind certainly does blow.

When augmenting a windpower system with photovoltaics, you'll also find that wiring is not at all complicated. If you anticipate problems with overcharging, by all means include a *photovoltaic* charge regulator along with any other regulating unit that may already be in the system. Fig. 7 shows how this might be done.

It should go without saying that the panels should be placed in unshaded areas. But if they can't be located to catch every bit of the early or late sun, don't worry too much, as the most productive hours are between 10 AM and 4 PM. (If the system is set up to track the sun, its power output will increase by about 40%.)

Locate the storage batteries in a sheltered area that's well ventilated (explosive gases are given off by active batteries) and protected from temperature extremes. The storage site—as well as the panels—should also be as close to your house as possible to limit the line loss that occurs when electricity travels through wire. It's a good idea to check the charge in the batteries regularly with a hydrometer and add water as needed.

Even though the panels themselves shouldn't require maintenance, it doesn't hurt to dust the covers every so often to let in as much sunshine as possible. It's also both easy and advisable to make a quick inspection of the connections.

Yes, photovoltaic power is here, and with a little inventive acquisition, a person may well be able to set up an affordable system today. Furthermore, the odds are that in a few years solar-electric homes sporting rooftop panels will have become commonplace.

ELECTRICITY FROM THE SUN

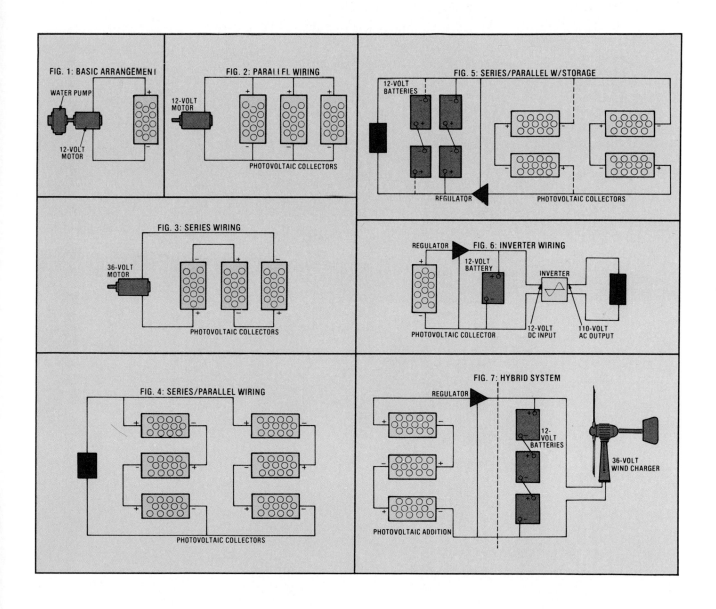

FIG. 1: BASIC ARRANGEMENT

WATER PUMP

12-VOLT MOTOR

FIG. 2: PARALLEL WIRING

12-VOLT MOTOR

PHOTOVOLTAIC COLLECTORS

FIG. 3: SERIES WIRING

36-VOLT MOTOR

PHOTOVOLTAIC COLLECTORS

FIG. 4: SERIES/PARALLEL WIRING

PHOTOVOLTAIC COLLECTORS

FIG. 5: SERIES/PARALLEL W/STORAGE

12-VOLT BATTERIES

REGULATOR

PHOTOVOLTAIC COLLECTORS

REGULATOR   FIG. 6: INVERTER WIRING

12-VOLT BATTERY

INVERTER

PHOTOVOLTAIC COLLECTOR   12-VOLT DC INPUT   110-VOLT AC OUTPUT

FIG. 7: HYBRID SYSTEM

REGULATOR

12-VOLT BATTERIES

36-VOLT WIND CHARGER

PHOTOVOLTAIC ADDITION

FUELING AROUND WITH ALCOHOL

Following immediately on the heels of the astronomical increase in the price of gasoline and other traditional motor vehicle fuels in the early 1970's, many people around the world began experimenting with less expensive alternatives to the nonrenewable petroleum products they had relied upon for decades.

Many folks turned to alcohol to power their automobile engines. Manufactured by fermenting and distilling vegetable matter—usually corn or a similar starchy grain—ethanol can be made by an individual in quantities to suit his or her own needs. By growing their own grain and fabricating their own stills, many people have used alcohol—either alone or in combination with traditional fuels in dual-function systems—to cut down on their trips to the gas pumps.

FUELING AROUND WITH ALCOHOL

Making Mash
The Grain's the Thing

The first step in the distilling process (once you have your equipment and raw materials, of course) is to make the mash. Successful ethanol production begins with a starch-rich concoction which converts, easily and thoroughly, to a yeast culture's favorite food: fermentable sugars. When the fungi find a generous supply of their preferred fare, they reward the distiller with a high rate of alcohol yield. A well-tested corn-based recipe is given on pages 64 and 65. While ethanol can be made from other starch products, tests show that this formula returns a maximum volume of fuel for minimal expense and effort.

Cutting corners will compromise the quality of the product and may even render it useless. You will find that a proven recipe, the right tools, and a few precautions will keep you from wasting time and money.

First, you should know how and why a mash happens. What is it that occurs in the vat to turn starch into sugar and sugar into alcohol? If you have a basic understanding of what's going on in your mash tub, you'll be less likely to make time- and money-wasting mistakes.

The addition of enzymes is the key step in the conversion of the starch to sugar. The enzymes act upon starch to change it into simple fermentable sugars such as maltose and glucose, a necessary process in the making of alcohol. Some home distillers will be using materials which don't need to go through this step because the substances already contain fermentable sugars. Some examples are fruit squeezings, molasses, and whey. But most ethanol manufacturers are likely to begin with starchy substances such as corn, wheat, or tubers and will have to convert the starches that these products contain.

There are several different ways to go about the conversion, but all of the various methods require the same substance, an enzyme called amylase. Amylase preparations are the biological catalysts that can break the bonds of the starch molecule to free the sugar molecules of which it is composed. This conversion of a substance from starch to sugar is known as the *saccharification* process. The enzymes that perform this necessary function are contained both in malted (or sprouted) grains and in commercially produced and purified enzyme solutions. Either of these enzyme sources will competently do the work of converting starch into sugar. And it's important to remember that microbes, which produce the enzymes, are alive, and thus require a specific, stable niche if they're to remain viable. If improper procedures are used during the starch to sugar conversion process—if, for example, the mash becomes contaminated or if heat levels are too high—the alcohol yield can be adversely affected. Correct and stringent sanitary procedures and careful attention to temperatures are necessary to prevent a lessening of the yield.

The recommended corn-mash recipe calls for the use of two different amylase enzymes. The first is a food-grade bacterial *alpha-amylase*, which is capable of rapidly reducing the viscosity of gelatinous starches and converting them into dextrins. Because it can be used in starch liquefaction processes (such as brewing) that employ high temperatures, the alpha-amylase is a valuable tool whenever heat is used in the gelatinization of a raw material such as corn.

The recipe also calls for a complicated glucoamylase—an inexpensive malt replacement

STILLS AND MASH, THE COMPLEAT COOK

developed by the neutral grain spirits industry—which actually breaks down the starch molecules, converts them into fermentable sugars, and denatures the mash.

Malting (sprouting) your own grain is an inexpensive method of producing enzymes without having to buy a commercial preparation (either cereal or bacterial), but it does entail a lot of work. All starchy (cereal) grains produce both alpha- and beta-amylase enzymes when moistened and allowed to sprout, and the method of sprouting to produce enzymes was used by home distillers and brewers for hundreds of years before commercially produced cereal and bacterial enzymes became available. While barley has been found to produce the greatest quantity of enzymes by volume—making it the most economical grain to use—corn, wheat, rye, and other grains can also be sprouted. The thing to remember is that undried, or "green", sprouted grain will not keep and has to be used immediately. Because of the time and work involved in sprouting grain and the additional labor that's necessary to dry a portion to use in malting subsequent batches (so that you won't have to go through the sprouting process with every single batch of alcohol you make), most folks find it far more cost-effective in the long run to buy commercially malted grain or to use the bacterial enzymes that have already been mentioned.

Just as enzymes are responsible for the conversion of starch to sugar, *yeast* converts sugar to alcohol. Yeast is a fungus that thrives in sugar solutions. When placed in a sweet liquid, the yeast "plant" goes through eleven separate conversion stages, secreting a complex of enzymes which are responsible for the production of alcohol from sugar. The whole process is called *zymolysis*, a term that refers to a procedure in which several enzymes and coenzymes actually perform the operation of removing an atom here and replacing it there. The result is the breakdown of sugar into carbon dioxide and alcohol.

The point of this entire process—as far as the yeast is concerned—is reproduction. As they multiply, the tiny plants just happen to create the by-products for which they are prized.

In fermenting solutions, carbon dioxide bubbles are produced that usually evaporate into the atmosphere, leaving only the ethanol and other liquids which make up the "beer". However, in large commercial distilling operations, provision is sometimes made to capture the carbon dioxide in separate holding tanks so it can be sold. Doing this makes the production of alcohol from vegetable matter an even more economically feasible proposition.

Because yeast is a living organism, the environment in which it lives will have a great effect on how efficiently it does its job. In the absence of air (oxygen in the water of the slurry), a far greater portion of the sugar goes off as alcohol and carbon dioxide. On the other hand, if the mash is aerated (exposed to the air, as by stirring), the yeast will produce more new yeast cells but fewer valuable "wastes". The idea is to leave the yeast undisturbed so it will give a greater alcohol yield.

Heavy metals and other chemicals can retard or stop yeast growth, which can also be adversely affected by heat or cold. A temperature of between 77° and 90°F should be maintained during fermentation.

FUELING AROUND WITH ALCOHOL

Making Mash
The Grain's the Thing (continued)

Yeast plants can't go on making alcohol and carbon dioxide forever. If they could, there would be no need for distilling to concentrate the ethanol. Yeast begins to become dormant when alcohol concentrations in the solution reach about 12% by volume, thus ending the sugar conversion process. It is at this point that distilling must begin.

Since the degree of alkalinity or acidity is crucial to such microorganisms as enzymes and yeast, it's quite important that you monitor the pH-level of your mash closely. The pH of a solution is simply a measure of acidity or alkalinity, expressed on a scale of 1 to 14, and an absolutely neutral mixture will have a pH of 7. Readings below 7 indicate acidity, and measurements above 7 indicate alkalinity.

At various points throughout the processes of mashing and fermenting, pH readings must be taken, and pH levels adjusted. Readings can easily be taken with pH test papers (litmus papers) which are available from laboratory supply outlets, swimming pool supply houses, and garden supply houses. The pH level can be adjusted by adding sulfuric (or citric or hydrochloric) acid to lower the reading, or agricultural lime to raise it. Most commercial enzymes, such as those used in the corn mash recipe, don't require any such adjustments.

Most grain mashes tend to be naturally acidic, while saccharide mashes (those made from fruit) will usually be alkaline. Either extreme must be adjusted after testing.

Fermentation is—in this case—simply the action of microscopic yeast fungi upon a sugar solution. The minute plants consume the sweet substance (which is either added to the mixture in the form of cane sugar, or "freed" from the corn by sprouting, or both) and give off carbon dioxide gas and alcohol.

When the yeast is added to the mash (the brew should be at a temperature between 70° and 90°F), fermentation begins. Any of a variety of yeasts can be used, from active dry yeast (which is usually available only in large quantities) to baker's yeast (which can be obtained from bakery supply houses or, perhaps, from your local baker) to the powdered dry or cool cake yeasts you can find in any supermarket.

All you need to do after your brew has cooled properly and you've added the yeast is sit back and wait for it to ferment. But remember . . . one of the most important details in preparing any mash is to keep your containers and implements clean. The fermentation process depends upon the action of the yeast alone. The presence of outside microorganisms—such as those found in the air all around us—can render a whole batch useless.

Therefore, it's very important to use bleached and scrubbed (and thoroughly rinsed) plastic or wooden vessels for every step in the mixing and fermenting process. (Metal drums can also be used, if they're lined with fiberglass or a plastic sack and made to seal.)

Be sure to cover the barrel to keep air from the mash. (Not everyone does this, but it does improve the efficiency of fermentation . . . after all, we're trying to produce an ideal batch of alcohol, not an ideal environment for yeast to reproduce in record numbers.) Many distillers equip the lid or cover with some type of air or fermentation lock that lets the fermentation-produced CO_2 out, while not allowing any air in.

STILLS AND MASH, THE COMPLEAT COOK

As has been pointed out, the mash should have cooled to a temperature of between 70° and 90°F before yeasting. After yeasting, fermentation begins, and the little plants start to reproduce. At the height of fermentation, the carbon dioxide gas given off during the process will actually create heat within the mash. The mixture will appear to boil at this stage, though of course that apparent "boiling" is brought about by escaping gas.

Since you must not allow the temperature of the mash to rise above 95°F (remember that yeast plants can be killed or rendered dormant by extremes of temperature), it's important that you monitor your batch and cool the mixture if you're preparing large enough batches of mash to produce a significant temperature change. While this can be done in a number of ways, the most efficient is to sink a cooling coil inside of the mash tub. It's mandatory for large-scale alcohol production (using vats of around a 1,000-gallon capacity), and it's also necessary for achieving successful results from the recipe given.

To make a cooling coil, just wind a 30-foot length of soft copper tubing around a large pipe (6 inches or greater in diameter), and add garden hose adapters at each end. Attach hoses to the tube, and drop the assembly into your cooking vat. When the temperature of the fermenting mash starts to get too high, simply run cool water through one hose, into the coil, and out again. It's as easy as that. Just be certain to monitor the temperature periodically. If you allow it to fall below 60°F, the yeast will become dormant, and the fermentation will stop.

As an alternative, you can lower the temperature by removing the cap from the fermentation tub and letting air cool the mash, but this will add a day or so to the fermenting process and (though it's unlikely) may allow harmful bacterial matter to get into the batch. Or, you can suspend plastic trash bags full of ice in the mash, but this technique is impossible to regulate, and it inevitably disturbs the mash.

Fermentation should be complete in two-and-a-half to three days, using the following recipe. Watch the action of the bubbles in the air trap closely after the third day (but don't open the containers). When carbon dioxide gas no longer "perks" through the vent hole, your brew is ready. (Be careful, though: Mash has been known to stop "working" for a day, only to start bubbling again with a vengeance.)

Another method of telling when your batch is ready is to wait for the "cap"—a layer of matter that will build on top of the mash—to drop to the bottom of the container. (Of course, this technique is best used with a transparent or translucent vessel.) When the mash has finished working, simply uncap the container, strain the mixture through a few pieces of washed burlap, and pour the resulting pure mash into whatever still you're planning to use. You're now ready to embark on your career as a distiller of fuel-grade alcohol.

CORN MASH RECIPE

First, shell, clean, and grind a bushel of corn (56 pounds) into a fine meal of about the size needed for livestock feed. Use a 3/16" screen on a hammer mill (or a similar grinder) to eliminate any large starch grains. However, do not grind the corn into a flour. If the grains are too small, it'll be very difficult to separate the solids from

FUELING AROUND WITH ALCOHOL

Making Mash
The Grain's the Thing (continued)

RAW MATERIAL	PREPARATION	ADDITIVES (ENZYMES)	PREBOIL
STARCHES			
WHEAT, CORN, RYE, BARLEY, MILO, RICE, CATTAILS	Grind to a fine meal using a 3/16″ screen on a hammermill; add 30 gal. water per bushel.	Add 3 spoons mash cooking powder* per bushel.	Raise temp. to 170°F for 15 min.; agitate vigorously.
PASTRY WASTE	Break apart, do not grind; add 30 gal. water per 55 lb.	As above.	As above.
POTATOES, CASSAVA (MANIOC), TARO	Slice, crush, or break apart; add 10 gal. water per 100 lb., or as little water as possible.	Add 5 spoons mash cooking powder* per 100 lb.	None.
SUGARS			
SUGAR BEETS, MANGEL-WURZELS, ARTICHOKE TUBERS	Slice or crush; add 10 gal. water per 100 lb., or as little as possible.	Acid may be added to beets to reach pH 5.0.	None.
SWEET SORGHUM, CANE, ARTICHOKE STALKS	Squeeze out juice.	None.	Raise temp. to 180°F for 10 min. to sterilize.
MOLASSES, SUGAR PRODUCTS	None.	Molasses from beets may need neutralization with acid.	If necessary, sterilize as above.
CHEESE WHEY	None.	None.	None.
CELLULOSE	Chop straw or soft material. Wood must be fine sawdust or treated with 400°F steam for 2 hr.	Add a 1% caustic solution; hold at 140°F for 3hr. to separate lignin.	Draw off lignin; neutralize.

STILLS AND MASH, THE COMPLEAT COOK

COOK	COOL DOWN	CULTURE	COMMENTS
Hold at rapid rolling boil for 30 min.	Cool with coil to 170°F; add 3 spoons mash cooking powder*; agitate for 30–60 min.	Reduce temp. to 90°F; add 6 spoons mash fermenting powder*; agitate for 10 min.; cover.	Results: 9% alcohol. Wheat, rye, and barley may cause foaming: Use No-Foam* or mix with cornmeal.
As above.	As above.	As above.	As above, plus remove oil (if content is high) before fermentation.
Raise temp. to 180°F for 30 min.; agitate vigorously.	None.	Reduce temp. to 90°F; add 10 spoons mash fermenting powder*; agitate 10 min.; cover.	Results: 9% alcohol.
Raise temp. to 190°F for 20 min.; agitate.	None.	Reduce temp. to 90°F; add yeast; agitate 10 min.; cover.	Results: 7% alcohol. Beets may require some molasses yeast food*.
None.	None.	Reduce temp. to 90°F; add water to make 18% sugar; add yeast; agitate 10 min.; cover.	Results: 9% alcohol. Molasses yeast food* may be added to increase yield.
None.	None.	As above.	Results: 9% alcohol. Use molasses yeast food* to insure proper yield. High NaCl content may interfere with fermentation.
Raise temp. to 210°F for 10 min. to sterilize.	Separate protein with NH₄OH; adjust pH to 5.0.	Reduce temp. to 90°F; Add *Kluyveromyces fragilis* or *Torula cremoris* yeast. Fermentation takes only 12 hr.	Results: 3% alcohol. Aeration may increase yield. Whey may be used as liquid with corn, but lactose must be added for a conversion.
Cook at 140°F for 4 hr. in 1% solution of Allcellulase*.	Remove sugar liquid.	Reduce temp. to 90°F; add brewer's yeast; agitate for 10 min.; cover.	Results: 2.5% alcohol. Acid hydrolysis is an alternative but expensive method.

* Available from Alltech, Inc., Dept. TMEN, 271 Gold Rush Rd., Lexington, Ky. 40503 Phone: 606/278-4358. (Note: Alltech's molasses yeast food is known as AYS.)

FUELING AROUND WITH ALCOHOL

Making Mash
The Grain's the Thing (continued)

the mash and will result in a loss of feed grain and a miserable mess inside your still.

Next, start with 30 gallons of water in your cooker, and then add the cornmeal slowly to prevent lumping. Once the meal is mixed in, stir in approximately one ounce of an alpha-amylase enzyme mixed in water, and bring the mixture up to 170°F (77°C). Hold the mash at this temperature for 15 minutes, stirring the brew vigorously throughout the process. Then bring the liquid to a rapid rolling boil and hold it there for 30 minutes more. Be particularly careful that the mash doesn't stick to the bottom of the cooker. (For batches larger than a bushel, it's a good idea to use an automatic agitator, which should spin at 30-45 RPM.)

Then using the cooling coil, bring the temperature of the mash down to 170°F (77°C), and add another ounce of enzyme (mixed in water). Keep the mixture at this temperature for 30 minutes, while you agitate it constantly.

Now, start cold water flowing through the cooling coil again to reduce the temperature to 90°F (32°C) as rapidly as possible. Once the mash has cooled, add two ounces of mash fermenting powder (see chart), stir the mash for 10 minutes, and then cover the tank.

While it's fermenting, the mash must be kept between 85° and 90°F (29-32°C). Consequently, you may need to cover the tank with wet burlap in hot weather and/or insulate it during the colder months. At this temperature, the mash will reach maturity in 2-1/2 to 3 days.

Finally, test the mash, and then test it again. At the beginning of fermentation, the specific gravity of the mash should be—when tested with a saccharometer—about 1.080 (8 to 12% alcohol potential), while by the end of the process it will have dropped to 1.007 or less (0 to 1% alcohol potential). Once the specific gravity has remained constant for 6 hours, you can be sure that the mash is ready for distillation. But to certify that complete conversion has been attained, both a standard starch test (using iodine) and a glucose test (using glucose test strips available at drug stores) must be negative.

This tried and proven recipe is corn-based, but in fact, the list of raw materials that can be used to make alcohol grows each day. Newcomers, such as mangel-wurzels (or fodder beets), Jerusalem artichokes, manioc, poplar trees, cellulose waste, and even cattails, have been added to the list of traditionals, which includes corn, sugarcane, potatoes, rice, and barley. (There are also peculiar—but potentially fruitful—food industry by-products such as waste pastry and stale tortilla chips.)

Despite the variety, every alcohol-producing raw material belongs to one of three groups: starches, sugars, or cellulose. And though the materials in each category are treated differently, the end product is always the same: glucose (or simple sugar), which yeast can easily convert to alcohol. The accompanying chart covers most of the major materials and how they should be prepared.

As a rule, sugar crops—such as sugar beets, sugarcane, and molasses—give a greater yield per acre than starch crops, because the material doesn't require conversion. Unfortunately, sugars don't store well. Processing includes squeezing the juice out of the stalks of plants or leaching it from their tubers. Whichever way you extract the sugar, be sure to sterilize the syrup to discourage contamination. Then, before you add the yeast, adjust the sugar concentration to 18%, using a saccharometer. In addition, yeast food should be introduced along with the yeast to increase alcohol production.

America "produces" 500 million tons of cellulose waste (such as wood chips and paper-processing by-products) annually. This waste, if properly handled, could yield almost 40 billion gallons of ethanol a year. However, cellulose is hard to break down because of a binding agent called lignin. Only in the last few years have researchers begun to develop economical methods of converting cellulose.

There are successful approaches based on both enzymatic and acidic conversion. For example, cellulase (the enzyme that converts cellulose to glucose) was isolated by the U.S. Army in 1945. Since then, that enzyme has been improved, although it is now fairly expensive.

Another method of converting tough cellulosic fiber involves forcing cellulose pulp (under high temperature and pressure) into a short, but intense, acid bath. The acid immediately converts the cellulose to glucose, but it must be removed quickly to avoid further processing and the destruction of the glucose. Researchers have developed very effective extrusion systems, but they're well beyond the pocketbook of the small-time operator.

The accompanying chart is meant to serve as a rough guideline to making mash . Once you start your own operation, you'll probably discover shortcuts that will allow you to use less heat and/or enzyme powder than the chart indicates. In addition, you'll need to look into proper mash testing and fermentation procedures, as well as into the recommended ways to handle and distribute your by-products. By studying time-tested methods and by following them meticulously, you can be assured of a favorable outcome. A successfully brewed mash is the result of good planning and careful preparation.

FUELING AROUND WITH ALCOHOL

A Couple of Homemade Stills

The growth in the popularity of alcohol as a fuel brought with it a search by farmsteaders and town dwellers alike for simple, cost-effective ways to "brew their own". The major start-up expense of any "homebrew" operation is the cost of the distilling equipment needed to produce fuel-grade alcohol, but the outlay can be held to a minimum by building your own moonshine maker.

There are a lot of tinkerers who might be interested in designing and building a still but who are undecided as to whether they actually want to produce alcohol fuel. They can legally assemble a still without a federal permit as long as the device is not used for distilling alcohol and has no mash introduced into it prior to the receipt of a permit.

The required license is issued by the U.S. Bureau of Alcohol, Tobacco, and Firearms, and no fee is charged. A special category of licensing has been established specifically for alcohol fuel producers. Anyone wishing to experiment with making ethanol fuel will be considered for an Alcohol Producers Permit. The procedure isn't complicated.

Producers of alcohol fuel may make and receive up to 10,000 proof gallons (one gallon of ethanol at 100 proof equals 1 proof gallon) per year *without* obtaining bonding. The bond requirements for plants producing more than that will be determined by the amount of alcohol distilled and received within a calendar year. Accurate records should be kept of the quantity and proof of spirits produced, the number of gallons on hand and received, the amount and types of materials added to render the alcohol unfit for beverage use (usually gasoline), the quantity of resulting fuel alcohol, and all dispositions of the spirits and/or fuel alcohol. As with any enterprise, applicable state regulations should be checked.

Both of the backyard stills presented here have proved to be effective alcohol factories, and either can be built easily and inexpensively by anyone with welding skills. The first distillery—simply a tank within a tank—makes a great "test bed" for various column designs and mash recipes. By using cast-aside materials for parts, anyone should be able to construct the mini-still for a reasonable cost, and the tiny "percolator" can turn out almost 1/2 gallon of 180-proof fuel per hour.

The major components of this baby distillery are two discarded water heater tanks (those from electric models are easiest to work with, and nongalvanized units don't give off noxious fumes during the cutting and welding process as do their coated cousins), some pipe for the column, filler, and drain, copper conduit for the condenser assembly, and a few assorted fittings and pieces of steel stock.

When choosing the tanks, make sure the vat-to-be is completely leak-free and is about 4" smaller in diameter (and 12" to 16" shorter in height) than is the firebox container. Before you cut the top off the larger tank and weld the small cauldron to it, mark the spot where the drainpipe will protrude through the firebox wall, and cut a 6" X 6" opening in the larger container's jacket at that point.

Follow the exploded drawing on page 69 to assemble the remainder of the still. The two-inch column can be filled with rolled-up nylon window screening, or—if you weld a perforated plate at

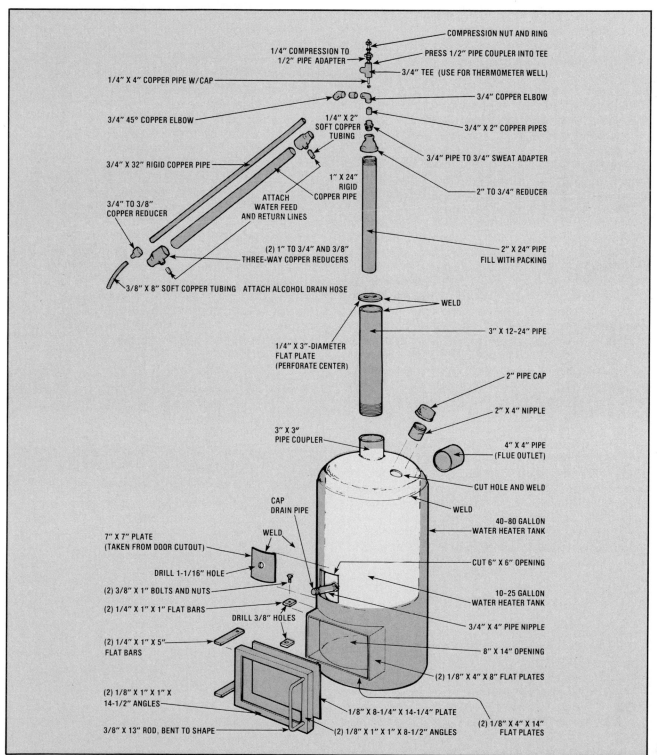

COMPRESSION NUT AND RING

1/4" COMPRESSION TO 1/2" PIPE ADAPTER

PRESS 1/2" PIPE COUPLER INTO TEE

3/4" TEE (USE FOR THERMOMETER WELL)

1/4" X 4" COPPER PIPE W/CAP

3/4" COPPER ELBOW

3/4" 45° COPPER ELBOW

1/4" X 2" SOFT COPPER TUBING

3/4" X 2" COPPER PIPES

3/4" X 32" RIGID COPPER PIPE

3/4" PIPE TO 3/4" SWEAT ADAPTER

1" X 24" RIGID COPPER PIPE

2" TO 3/4" REDUCER

ATTACH WATER FEED AND RETURN LINES

3/4" TO 3/8" COPPER REDUCER

(2) 1" TO 3/4" AND 3/8" THREE-WAY COPPER REDUCERS

2" X 24" PIPE FILL WITH PACKING

3/8" X 8" SOFT COPPER TUBING   ATTACH ALCOHOL DRAIN HOSE

WELD

1/4" X 3"-DIAMETER FLAT PLATE (PERFORATE CENTER)

3" X 12-24" PIPE

2" PIPE CAP

2" X 4" NIPPLE

3" X 3" PIPE COUPLER

4" X 4" PIPE (FLUE OUTLET)

CUT HOLE AND WELD

WELD

CAP DRAIN PIPE

40-80 GALLON WATER HEATER TANK

7" X 7" PLATE (TAKEN FROM DOOR CUTOUT)

WELD

CUT 6" X 6" OPENING

DRILL 1-1/16" HOLE

(2) 3/8" X 1" BOLTS AND NUTS

10-25 GALLON WATER HEATER TANK

(2) 1/4" X 1" X 1" FLAT BARS

3/4" X 4" PIPE NIPPLE

DRILL 3/8" HOLES

(2) 1/4" X 1" X 5" FLAT BARS

8" X 14" OPENING

(2) 1/8" X 4" X 8" FLAT PLATES

(2) 1/8" X 1" X 1" X 14-1/2" ANGLES

1/8" X 8-1/4" X 14-1/4" PLATE

3/8" X 13" ROD, BENT TO SHAPE

(2) 1/8" X 1" X 1" X 8-1/2" ANGLES

(2) 1/8" X 4" X 14" FLAT PLATES

FUELING AROUND WITH ALCOHOL

A Couple of Homemade Stills (continued)

the bottom of the tube—it can be packed with short sections of 1/2″ copper conduit, loosely woven rustproof metal scrubbing pads, or commercial packing. The condenser is nothing more than a conduit set within a larger tube that serves as a water jacket. (To keep track of column temperature for testing purposes, replace the outlet elbow for the vapors with a tee fitting, which can then be used to serve as a thermometer holder.

The second design—a four-inch column model—is capable of producing about two gallons per hour of 90% pure ethanol fuel. Naturally, because this model is larger and somewhat more sophisticated than the two-inch column still, its cost will be greater, but the entire assembly can probably be built in about 30 hours.

Although the illustration on page 71 is largely self-explanatory, there are a few fine points that should be mentioned. The 40- to 80-gallon tank should, of course, be leak-free and preferably nongalvanized, and all its unused holes must be plugged. Since the column will have to be filled with pall rings (or some other type of loose material), it will also be necessary to insert a drilled packing support plate in the pipe—at its bottom flange—to prevent the packing from falling into the mash vat.

Both the internal heat exchanger and the condenser are simply lengths of 1/4″ OD copper tubing, wound into 2″-diameter coils and held in place—within their respective columns—by compression-to-pipe adapters, which are themselves fastened to pipe couplers welded to the tube's walls. (The coils can be formed by wrapping the soft tube around a 1-1/2″ thick section of pipe or wooden dowel.) Since the amount of water flow controls the critical temperature within the packed tower, the rush of cooling liquid—to the column, at least—should be regulated by a needle valve on the supply side, although separate controls and lines to the heat exchanger and condenser (using a tee and a single water return hose) may be more practical. Remember, too, that it's possible to install a thermometer well at the top of the column, in order to keep vapor temperatures within the desirable 175–180°F range.

Though neither of these two stills will—by itself—be capable of producing all the fuel an average American family consumes, either one certainly would be able to supply a motorcycle, home-generating unit, garden tractor, or any one of countless other pieces of normally gasoline-powered equipment in use on homesteads and in communities across the country today. Just bear in mind that anyone experimenting with producing ethanol or distilling alcohol should operate within the guidelines set by the U.S. Bureau of Alcohol, Tobacco, and Firearms.

STILLS AND MASH, THE COMPLEAT COOK

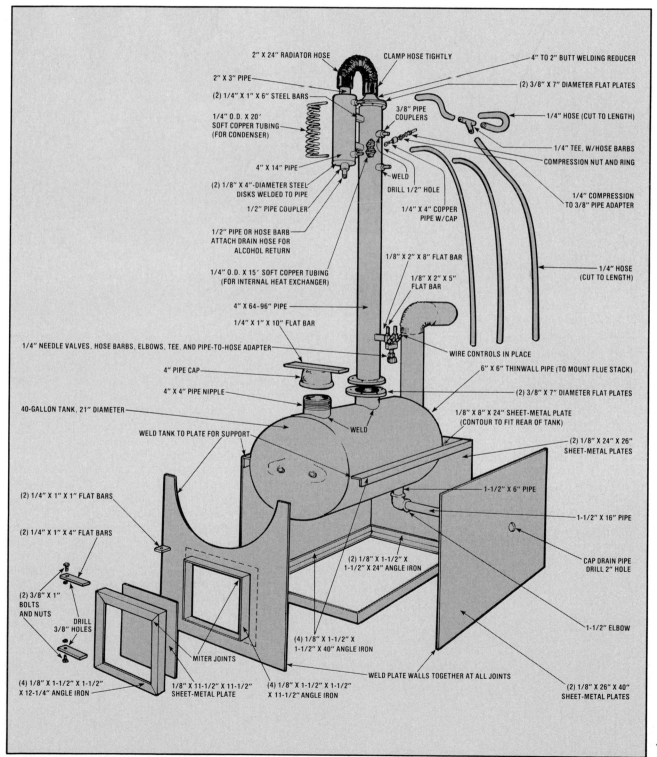

2" X 24" RADIATOR HOSE

CLAMP HOSE TIGHTLY

4" TO 2" BUTT WELDING REDUCER

2" X 3" PIPE

(2) 3/8" X 7" DIAMETER FLAT PLATES

(2) 1/4" X 1" X 6" STEEL BARS

3/8" PIPE COUPLERS

1/4" HOSE (CUT TO LENGTH)

1/4" O.D. X 20'
SOFT COPPER TUBING
(FOR CONDENSER)

1/4" TEE, W/HOSE BARBS

COMPRESSION NUT AND RING

4" X 14" PIPE

WELD

(2) 1/8" X 4"-DIAMETER STEEL
DISKS WELDED TO PIPE

DRILL 1/2" HOLE

1/4" COMPRESSION
TO 3/8" PIPE ADAPTER

1/2" PIPE COUPLER

1/4" X 4" COPPER
PIPE W/CAP

1/2" PIPE OR HOSE BARB
ATTACH DRAIN HOSE FOR
ALCOHOL RETURN

1/8" X 2" X 8" FLAT BAR

1/4" O.D. X 15' SOFT COPPER TUBING
(FOR INTERNAL HEAT EXCHANGER)

1/8" X 2" X 5"
FLAT BAR

1/4" HOSE
(CUT TO LENGTH)

4" X 64-96" PIPE

1/4" X 1" X 10" FLAT BAR

1/4" NEEDLE VALVES, HOSE BARBS, ELBOWS, TEE, AND PIPE-TO-HOSE ADAPTER

WIRE CONTROLS IN PLACE

4" PIPE CAP

6" X 6" THINWALL PIPE (TO MOUNT FLUE STACK)

4" X 4" PIPE NIPPLE

(2) 3/8" X 7" DIAMETER FLAT PLATES

40-GALLON TANK, 21" DIAMETER

1/8" X 8" X 24" SHEET-METAL PLATE
(CONTOUR TO FIT REAR OF TANK)

WELD

WELD TANK TO PLATE FOR SUPPORT

(2) 1/8" X 24" X 26"
SHEET-METAL PLATES

1-1/2" X 6" PIPE

(2) 1/4" X 1" X 1" FLAT BARS

1-1/2" X 16" PIPE

(2) 1/4" X 1" X 4" FLAT BARS

(2) 3/8" X 1"
BOLTS
AND NUTS

DRILL
3/8"
HOLES

(2) 1/8" X 1-1/2" X
1-1/2" X 24" ANGLE IRON

CAP DRAIN PIPE
DRILL 2" HOLE

1-1/2" ELBOW

MITER JOINTS

(4) 1/8" X 1-1/2" X
1-1/2" X 40" ANGLE IRON

(4) 1/8" X 1-1/2" X 1-1/2"
X 12-1/4" ANGLE IRON

1/8" X 11-1/2" X 11-1/2"
SHEET-METAL PLATE

(4) 1/8" X 1-1/2" X 1-1/2"
X 11-1/2" ANGLE IRON

WELD PLATE WALLS TOGETHER AT ALL JOINTS

(2) 1/8" X 26" X 40"
SHEET-METAL PLATES

FUELING AROUND WITH ALCOHOL

**The "Cornfed Cruiser"
Is No Gas "Hog"**

For various reasons, including their moderate consumption of fuel and their maneuverability, motorcycles are the preferred mode of transportation by many. But when the rising cost of gasoline increased the cost of operating almost any vehicle, some drivers began to consider the use of other, less costly fuels. In a demonstration of the feasibility of using alternatives to fossil fuels, a 1979 Harley Davidson Sportster was converted to burn ethanol, which can be produced relatively inexpensively in a homemade still.

Once converted, the air-cooled, 61-cubic-inch engine—normally characterized by both power and vibration as a result of its long-stroke, twin-cylinder design—immediately became more "civilized", while apparently delivering greater amounts of torque. This possibly is because the slow-burning, denser-than-gas alcohol distributes pressure evenly on the piston head. The only noticeable disadvantages are a slight reduction in top speed and a 7.8% decrease in mileage, but these were well compensated for by a smoother, more comfortable ride.

A lower engine temperature and less oil consumption were two results of burning the home-grown fuel. Because ethanol has a high latent heat of vaporization compared with that of gasoline, it absorbs a good deal more warmth from the engine's manifold and cylinder heads—in the process of changing from a liquid to a vapor

This Harley-Davidson Sportster was converted to burn alcohol rather than gasoline. To facilitate cold-weather starts, a squirt of gasoline from the thumb-operated pump is used to prime the engine, and a copper preheater warms the alcohol before it enters the carburetor.

MAKING THINGS GO WITH ETHANOL

state—than does its more conventional fuel counterpart. The result is a reduced engine operating temperature, a welcome bonus which can allow longer periods between oil changes, especially in an air-cooled powerplant.

The process of converting the bike to burn 185-proof fuel was relatively simple. First, the main jet on the carburetor was enlarged 35% from its original diameter by using a small jeweler's bit and a pin vise. Then the fuel port in the idle circuit was reamed to about 45% larger than normal. In addition, the motorcycle's response was markedly improved when the length of the accelerator pump stroke was decreased: When the throttle was cracked open, the amount of additional fuel injected into the carb throat was slightly reduced, thereby diminishing the liquid's tendency to form large droplets rather than to vaporize.

Further testing indicated that advancing the ignition timing—a modification that's normally a matter of course for any alcohol-powered engine, because of the fuel's high octane rating—had such a minor effect on the vehicle's performance, that it's best to leave this setting at factory specifications.

The remaining modifications to the bike were made to overcome the ethanol's resistance to vaporization in cold weather, and these would be necessary only if the cycle were to be used in temperatures below 45°F. First, a homebuilt canister was fastened to the frame of the motorcycle, and inside that was mounted a plunger-type plant waterer. A short section of 1/8″ model airplane fuel line was run from the outlet nozzle of the thumb-operated squirter to a needle valve, and a length of thin copper tubing was routed from this control to a small hole drilled in the wall of the carburetor air inlet. Precautions were taken so that neither the opening nor the tube inside it interfered with the movement of the choke plate.

When the compact container is filled with gasoline, a pump or two on the plunger will provide a starting "prime" for the engine. The amount of boost can be regulated with the needle valve.

The second modification, which enhanced the engine's operation in cold weather (but had little effect at other times), was a fuel preheater. It was made by wrapping a copper strap (about 1/2″ wide and 3″ long) completely around the middle of a short piece of copper tubing that had an outside diameter compatible with the inside dimensions of the cycle's fuel line. After the strap and tube were soldered together, the alcohol fuel line was cut in two and spliced to the two ends of the preheater. The free end of the copper strap was bolted to a "hot spot" on the nearby engine, which allowed sufficient thermal energy to be transferred through the device to warm the fuel.

The motorcycle was tested for over a year to determine the long-term effects of the alcohol on the bike's fuel system. After that period, there was no noticeable deterioration other than a slight swelling of the O-ring seals, which was found when the carburetor was disassembled for inspection. However, the rings returned to original size when they dried out. And, though a small amount of white, powdery residue was found in the carburetor float bowl and in the fuel tank, there seems to be little problem with using ethanol. In fact, this two-wheeled transporter adapted so readily to "homegrown energy" that it may never see another drop of gasoline.

FUELING AROUND WITH ALCOHOL

A Dual-fueled Van—
Two Fuels Are Better Than One

The size of the vehicle—be it a motorcycle, an automobile, a piece of gardening equipment, or whatever—seems to be no problem when using ethanol as a fuel. And, while an alcohol-fueled eighteen-wheeler may be only a gleam in a designer's eye, there's no reason why you can't harness this fuel to a workhorse like this 1977 Chevy 1/2-ton van equipped with a 250-cubic inch "six", an automatic transmission, and a "flip of the switch", genuinely efficient dual-fuel system.

As you might imagine, this dual-fuel system utilizes two carburetors. The van's original Rochester downdraft model was replaced with a pair of side-draft Carter YH carburetors (an older automotive model that's now widely used in marine applications) mounted on a homebuilt "Siamese" Y manifold that was fabricated from muffler tubing and flanges. The front "fuel mixer" is relatively unchanged and runs on gasoline, while the rear atomizer has been modified to suit the requirements of ethanol. The mechanisms are individually controlled by a solenoid-operated, selective sliding throttle linkage made from some flat stock and a 4″ door hinge.

In addition, in order to improve vaporization, the air entering the ethanol carburetor was preheated (rather than the fuel) from ambient temperature to 170°F. Also, advantage was taken of the liquid's high "octane" rating by automatically advancing the ignition timing when burning alcohol. Spark "progression" was carefully controlled in the gasoline mode to guarantee knock-free performance.

The conversion to the twin carburetor system used in the van was somewhat involved. Since the goal was to improve both fuel economy and vehicle drivability, horizontal-style carburetors were used, not only because they offer the convenience of dual-fuel capability, but also because their side-draft design affords the least restrictive fuel flow from the float bowl to the "booster" venturi area within the mechanism's throat. Though some of the internal passageways still needed modification, conventional downdraft carbs would require reworking, too, but might not perform as well because of their "serpentine" fuel-feed galleries.

Before any modifications are begun, the carburetor that is to be alcohol fueled must be dismantled and thoroughly cleaned with solvent. This preparatory operation includes the removal of the press-fit plugs that seal the factory-drilled

This Chevrolet van has been converted to run smoothly, efficiently and economically on both alcohol and gasoline.

fuel passageways from the outside. Flat and needle-nosed pliers, a screwdriver, a wire gauge set, and a complete assortment of drill bits (numbered 1 to 80, plus one 19/64″ version) are needed to perform the surgery.

The first step involves enlarging the main jet from .080″ to .104″, for a total diameter increase of 30%. Then increase the diameter of the accelerator pump discharge nozzle by 20%, from .025″ to .030″. Next, to improve the engine's transition from idle to cruising speeds, expand the idle transfer slot (part of a circuit that supplies nearly all of the necessary air/fuel mixture during normal driving at speeds below 40 MPH) from .020″ to .059″, and drill two .043″ holes in line at the "upstream" end of the slot to further enhance smooth performance during "off idle" operation. Finally, enlarge the idle screw passageway leading from the needle's tapered tip to the throttle bore—to provide a wider range of adjustment—from .066″ to .076″, and raise the float height by 3/16″.

The second phase of the modification procedure includes drilling a "sight hole" in the side of the float bowl so that the port's lower edge is about 1/10″ below the dripping orifice, visually setting the float to eliminate the leakage, and tapping the inspection hole so a threaded brass plug can be installed after final adjustments are made. In addition, the idle circuit passageways

The alcohol carburetor's ability to vaporize its fuel is greatly improved by preheating the intake air with the van's existing hot water system . . . and the throttle linkage selects the carburetor to be used.

FUELING AROUND WITH ALCOHOL

A Dual-fueled Van—
Two Fuels Are Better Than One (continued)

within the carburetor should be drilled out to .149", and the idle fuel restrictors and other potential problem areas (such as gasket perforations and casting burrs) should be either altered or removed.

At the same time, increase the idle fuel jet in the main well from .024" to .039", and enlarge the idle air bleed by 62% to .073". These alterations are made to increase both the amount and velocity of air/fuel mixture delivered to the engine from idle to 20% of full throttle position.

Other changes are as follows: Both the primary and secondary main fuel wells need to be enlarged to .300" in diameter. These reservoirs hold a supply of fuel "in waiting" between the main jet and the venturi supply tubes, for use the instant the throttle is cracked open. This provides an extra-rich fuel mixture to the engine as soon as it's needed. Then add a main fuel well air bleed, by drilling a .025" hole obliquely from a point near the top of the secondary main fuel well directly into the carburetor throat ahead of the booster venturi section of the carburetor. This passageway—a standard feature on newer carbs—supplies positive air pressure to the fuel wells to insure faster liquid movement and better response.

Also, if the original booster distribution jet has a restrictive central corridor that doesn't allow for an air mix, a new air/fuel mixture tube must be fabricated. In this case, a .128"-inside-diameter brass tube, with the same overall length and outside diameter as the jet, can be "slash cut" on one end at an angle of about 10°. Then cross-drill four .040" holes approximately one inch from this angled outlet end of the pipe. After removing the old jet with a 1/8" drill, press the new fuel/air mixing assembly into place and hold it fast with a drop of thread-locking fluid. Such a combination booster-and-blending jet serves to atomize the fuel quite effectively.

Individual throttle control for each carburetor is accomplished inexpensively by adapting a 4" door hinge—and some spare parts—to do "double duty". The hinge pin is removed, eventually to be replaced with a similar rod about 2" longer, with a 1/16" hole drilled at one end and a knurled head at the other. Then the three "knuckles" of an extra hinge plate should be ground down at their shoulders to allow about 1/4" of lateral play in the component once it is positioned "around" the main hinge (see right-hand photo on page 75).

Next cut the triple-knuckled plate from the hinge set into three sections—each having one "loop"—and permanently fasten its two-cylindered mate to the top of the Y manifold, in an upright postition. At this point, weld three control "arms" onto their respective hinge sections, and drill holes through each to accept throttle rods. Then reassemble the linkage using the long pin as a pivot. (The extra length is added to accommodate the third hinge plate and the spring at one end—held fast with a cotter pin—which lets the plates slide sideways and function without binding.)

Finally, an "antidieseling" solenoid from a GM V-8 should be installed to automatically move the "master" plate right or left, depending on which carburetor is to be controlled.

Happily, the apparatus functions well in both theory and practice: When the solenoid shaft extends to the left, and the accelerator is depressed, the center plate pushes down on the hinge section that activates the alcohol car-

MAKING THINGS GO WITH ETHANOL

buretor. At the flip of a switch, the solenoid arm returns to the right—along with the master plate—and the gasoline carb is ready for action.

With this setup, it's not necessary to incorporate individual fuel valves for the "dormant" carburetor. The unused mechanism merely rests in its "idle" position, atomizing no fuel, while the other carb functions normally. This occurs because such a great volume of air is flowing through the in-use device that there's not sufficient "draw" to pull fuel from the inactive one. In addition, there's a less obvious benefit to this arrangement. When the engine is idling (in either fuel mode), both the carburetors—which, of course, "draft" equally at idle—can be adjusted to provide nearly any gasoline/alcohol ratio desired. This feature allows the engine to operate very smoothly on low-proof ethanol at both idle and starting speeds, thanks to the addition of a small petrol "boost".

Warming the alcohol carburetor's air supply notably improves the carb's ability to vaporize the stubbornly liquid fuel. So you might adapt an automobile heater core and plenum assembly to function as an "atmosphere" preheating canister. First, cut a circular hole in one side of the box, to correspond with the mounting flange of an aftermarket air-cleaner unit. Then rivet a threaded crossbar to the inside surface of the chamber to provide a mount for the rod that holds the filter in place. The "hot box" is finished by fitting and fastening a backing plate to the rear of the container and ducting a flexible plastic feed tube from this surface to the carburetor intake.

After the device has been mounted with two aluminum brackets, and the new heat exchanger has been connected to the vehicle's existing hot water system, a consistent supply of 170-175° air will be available at any time, once the engine reaches operating temperature. The original air filter hookup must be reconnected to the gasoline side of the system, demanding nothing but a slight relocation.

Rather than rely on a control cable, you can regulate the ignition timing by using equipment that is already on the vehicle. After advancing the initial timing 12 degrees over the factory specification (or to a point at which it seems to suit—at idle—both gasoline and alcohol fuels), tie the distributor's advance diaphragm control hose directly into the ported vacuum fitting on the ethanol carburetor. Since this connection functions only in the "off idle" position, the proper timing adjustments are automatically made—through both the vacuum and centrifugal advance systems—when the accelerator pedal is depressed.

The vacuum fitting on the gasoline carburetor is plugged, so ignition timing in the "petrol" mode is entirely reliant upon the centrifugal advance mechanism in the distributor.

On the van in which this dual fuel setup was first installed, the twin carburetor assembly proved reliable, practical, and remarkably flexible, burning both gasoline and high- or low-grade alcohol fuel (right down to 170-proof "juice") with nary a cough nor a sputter.

FUELING AROUND WITH ALCOHOL

"Keep on Truckin' " with Alcohol

There's certainly nothing distinctive about an old pickup truck, but when such a vehicle is outfitted so it can bypass the gas pumps indefinitely, it becomes something worth writing home about. Converting a truck to do just that is simply a matter of making the appropriate adjustments and alterations to allow the truck to run on homemade alcohol.

Such a "conversion" isn't at all difficult, either. In fact, it can be done in less than two hours on just about any vehicle manufactured today, using tools found in almost anyone's workshop.

There's no reason for alcohol not to be used as motor fuel. Some of the earliest "horseless carriages" ran on it exclusively, and even in modern times, aircraft and racing cars have taken advantage of the fuel's benefits: [1] Alcohol burns clean. [2] The distilled fuel also acts as a cleaning agent within the engine. [3] An alcohol-burning engine tends to operate at slightly cooler temperatures than does its gasoline-powered counterpart.

Even aside from these mechanical benefits, there are other less obvious advantages to ethanol fuel, one of the most important being the fact that it's not dangerously volatile, as is gasoline. Other positive points include the fact that a 200-proof "juice" isn't necessary and the fact that—after obtaining the appropriate government permits—a vehicle's owner can manufacture alcohol at home.

Begin the job of converting the engine by gathering up all the tools and hardware needed to complete the task. In most cases, a screwdriver, a pair of needle-nosed pliers, a putty knife, a set of assorted end wrenches, a pair of locking pliers, and a power drill—with bits ranging in size from .0635″ (No. 52) to .0890″ (No. 43)—are all that a person needs. To make the alteration easier, though, you might want to refer to a Motor, Chilton, or Glenn auto repair manual for exploded illustrations to guide you through the necessary carburetor disassembly. An alternative would be to purchase a carb rebuilding kit for your particular make and model. Such a kit can provide not only a working diagram but also a supply of gaskets and other parts that may get damaged during the strip-down process. Finally, on nearly all carburetors there is a removable main metering jet, and you'll probably want to purchase several of these from your automobile dealer so you can easily convert your vehicle's engine back to gasoline fuel if the need arises.

After these preliminary steps, remove the carburetor air filter housing—and all of its hoses, tubes, and other paraphernalia—from the engine. Next, disconnect the throttle linkage from the carburetor, and if your automobile is so equipped, any choke linkage rods that aren't self-contained on the carb body. Older vehicles might use a manual choke, and if this is true of your car or truck, remove the control cable and tie it out of the way.

Unscrew the fuel line from the carburetor inlet fitting and remove any other hoses that fasten to the carb, including vacuum and other air control lines. If you're not quite sure that you will remember exactly where all these hoses belong, it would be smart to label them and their fittings when you take the hoses off.

When the carburetor is completely free from all external attachments, remove it from the manifold. Single-barrel units usually have only two fastening nuts or bolts, while two- or four-barrel models use four-point mounts. Once the carburetor is off the engine, drain the gasoline from it by turning it upside down. If it's covered with grime, take the time to clean off the assembly with an automotive degreasing solvent other than a carburetor cleaner, which would deteriorate rubber parts.

In order to use alcohol fuel in an engine designed to burn gasoline, it's necessary to enlarge the opening in the carburetor's main jet (or jets,

if your carb is a multithroat model). Start by removing the air horn from the float bowl (photo at left). In most cases there will be a choke step-down linkage rod—and possibly some other mechanical connection—between the two components. If possible, disconnect these before unthreading the air horn's fastening screws.

Next, locate the main jet (photo at right). Some carburetors—such as the unit pictured—have the jet installed in a main well support (a tower-

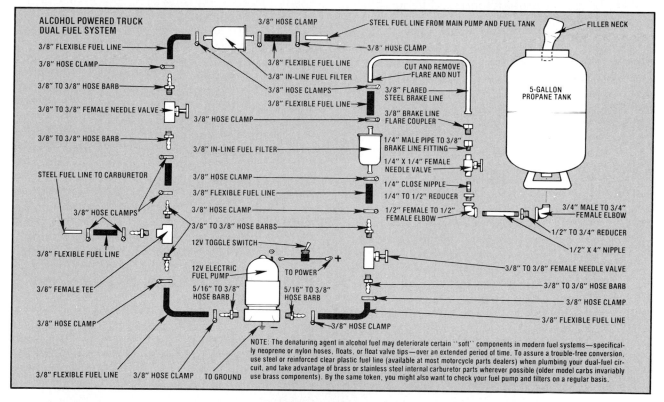

NOTE: The denaturing agent in alcohol fuel may deteriorate certain "soft" components in modern fuel systems—specifically neoprene or nylon hoses, floats, or float valve tips—over an extended period of time. To assure a trouble-free conversion, use steel or reinforced clear plastic fuel line (available at most motorcycle parts dealers) when plumbing your dual-fuel circuit, and take advantage of brass or stainless steel internal carburetor parts wherever possible (older model carbs invariably use brass components). By the same token, you might also want to check your fuel pump and filters on a regular basis.

In order to enlarge the opening of the carburetor's main jet, the air horn must be removed from the float bowl. Then the float assembly is taken off and the brass main jet is unscrewed.

FUELING AROUND WITH ALCOHOL

"Keep on Truckin' "
with Alcohol (continued)

like mount fastened to the air horn), while others mount the metering device directly in the float bowl body. In any case, there shouldn't be any trouble identifying the removable main jet: It's a round brass fitting—with a hole in its center and a slot in its top—that threads into place.

Now remove the float assembly, unscrew the jet, and measure the diameter of its central orifice. The simplest way to do this is to find a drill bit that fits snugly in the hole and then determine the size of the bit by matching its drill number to its diameter—in thousandths of an inch—by using a conversion chart available at your local hardware store or in a machinist's handbook.

Once you've determined the "normal" size of your gasoline jet's orifice, prepare to increase that dimension by about 40%. Remember that this isn't a fixed percentage for every engine, and you might have to drill several different jets—in progressive increments above and below that figure—and try them out by actually running the vehicle to see how they work.

If the orifice is too small, it won't allow enough liquid to enter the system, and the engine will backfire and miss. (It may also burn valves if left in such a lean condition for an extended period of time.) On the other hand, if the jet is over-enlarged, the mixture will be too rich, and you'll waste fuel.

Be sure to hold the jet with locking pliers while you carefully bore out its central hole and if possible, use part of the carburetor itself as a "mount" when you drill. If you do this, carefully clean any brass residue out of the carburetor and its components.

Some carbs will need additional idle-circuit enlargement if the engine is to run properly. To accomplish this, remove the idle-mixture screw and drill into the orifice with a bit that's slightly larger than the original hole. Be cautioned, however, that this alteration doesn't apply to all types of carburetors. It would be best to install the carburetor with only the main jet enlarged and try it out before drilling the fixed idle circuit.

The final change to be made in the carburetor is to shim the idle-mixture screw spring with a couple of small lock washers. This allows the threaded metering device to be drawn out farther than normal without danger of its vibrating loose. (Tighten the idle-speed screw by about one and one-half turns at this time.)

With the modifications completed, replace the main jet in the carb, install the float, and reassemble the carburetor. Position a fresh gasket on the manifold (make sure both metal surfaces are clean) and bolt the carb assembly in place.

At this time, if you choose, you might want to rig up a dual-fuel system. This will allow the use of either alcohol or gasoline (with a bit of tinkering involved in the switch-over procedure), and it will entail only the installation of a second fuel tank, some additional plumbing, and an electric fuel pump.

Start by looking your vehicle over and deciding where you want to put your extra tank. On a pickup, the container should fit perfectly between the cab and the rear fender, but on a passenger car, it might have to be mounted next to the regular gasoline tank (and away from the hot-when-running exhaust system, of course). The vessel itself can be anything from a recycled propane tank to a fuel tank from a small car, but whatever you choose, be certain that the container is leak-free and mounted safely.

MAKING THINGS GO WITH ETHANOL

Once the vehicle is equipped with a dual-fuel capability (one hookup is illustrated in the accompanying diagram), you can begin to attune the vehicle to its new "feed". Reattach the throttle and choke linkages and any hoses you might have removed in the conversion process, then drain the gas tank and fill it with alcohol. You can also fill the auxiliary container with gasoline at this time. Shut off the valve that supplies the new secondary fuel pump, open the alcohol control valve, and start 'er up!

It'll take a few seconds to fill the empty float bowl, but the engine should soon start and run at a fast idle. Slowly decrease the RPM of the engine by backing off on the idle-speed screw until the engine wants to stall. At the same time, it may be necessary to gradually adjust the idle-mixture screw—by one-quarter-turn increments —till the engine's "roughness" evens out. Eventually, the vehicle should idle nicely, though it may run slightly faster than it did before.

Now replace the air cleaner housing and take the vehicle for a drive. It should perform normally. After driving about ten miles at cruising speeds, you should remove the spark plugs and check their electrodes. If the tips are covered with a white coating, the combustion chamber is getting too hot. Dismantle the carburetor again and enlarge the jet by one drill size (remember, the lower the number, the larger the bit). Test the car again and recheck the plugs. They should be covered with an even, light tan coating.

Remember that every engine is different, and as such, each will require you to "fiddle" with it a bit before it will operate normally. If your alcohol-powered convert doesn't perform satisfactorily, try these remedies. First, advance the timing several degrees by turning the distributor housing opposite to the direction in which the rotor spins. (Don't overdo it, or the engine will "ping" with preignition.) You can also disconnect the vacuum advance line to the distributor and plug it with a screw or a ball bearing. This will prevent a too rapid spark advance. You might try closing the gap in the spark plugs by .004" to .006", too. If you care to get more involved, raising the compression ratio—either by simply "milling" the head or by installing high-compression pistons —will improve both engine performance and fuel economy.

Altering your car's original carburetor may, of course, prove impractical because of the unit's particular design features. The best solution in such a case would be to buy a rebuilt carburetor from an older model. The bolt patterns on most manifolds haven't changed for a decade of more, and nearly all carburetors made in the 1960's can be converted without difficulty.

The pickup truck described earlier ran well on alcoohol fuel for about three years before being converted to utilize gas generated from wood scraps. It started easily in the morning and ran smoothly when it reached operating temperature (a manual choke aided in the "warming up" process). Fuel economy wasn't quite as good in the alcohol mode, but was close to normal, and power and acceleration were unaffected.

It's important to remember that there is still much to learn and that every engine is an individual, so keep an eye on your converted engine and be ready to spot a problem before it gets serious. Also, before driving an alcohol-powered vehicle on public roads, you should check any applicable motor vehicle laws.

FUELING AROUND WITH ALCOHOL

Getting Alcohol Going in the Winter

Because it is less dangerously volatile than gasoline, alcohol has the advantage of being a relatively nonexplosive fuel. This advantage, however, can become a minor problem during the times of year when the temperature is below 45°F, because the carbureted ethanol fuel has a tendency to remain in liquid form as it passes through the engine's intake manifold rather than to turn into a mistlike vapor as it should.

But there is a low-cost, easy answer to this problem. Merely inject a fine stream of gasoline—taken from the auxiliary fuel supply—directly down the carburetor throat. The minute spray is adequate to start the engine and keep it running till it's warm enough to utilize pure alcohol fuel.

To construct your own injector, begin by purchasing (or otherwise locating) all the plumbing hardware pointed out in the accompanying illustration. Then gather up an assortment of end wrenches, a screwdriver, a power drill with a 7/16″ bit, a soldering iron, a small tubing cutter, and some plumber's pipe-joint tape.

When everything's together, remove the air cleaner lid from its housing and determine the best place to position the "squirt" nozzle. Remember that you want the sprayer to shoot directly down into the throat of the carburetor, and that the movable choke plate should not interfere with the injector pipette.

Now drill a 7/16″ hole in the air cleaner cover at this spot. Cut a short length of 1/4″ copper tubing, and fasten the needle valve to the elbow as in our illustration.

Finally, insert your copper tube into the opposite end of the elbow and attach the entire assembly to the air filter cap.

With this "nozzle holder" in place, you can now solder a brass metering jet to its tip. Select a jet that will fit into the end of the pipe, and if possible, choose one with a tiny orifice (a hole about .020″ is fine). If you've picked up a nozzle with too large an opening, just solder it partially shut by inserting a thin needle into the opening and removing it after the solder has set.

Now place the air filter lid on top of its housing, bolt the cap into position, and complete the plumbing circuit as depicted in the illustration. Remember that this system is designed to be used in conjunction with a dual-fuel setup, so it's merely a matter of replacing the circuitry between the auxiliary fuel pump and the main fuel line, and installing the combination of parts detailed here.

Once this cold-start system has been installed, you'll find it easy to operate. Just before cranking the engine, open the choke (if it's manually controlled) and flip on the switch that activates the secondary gasoline pump for a few seconds. Then shut it off, close the choke as you would normally, and fire the engine. It should start immediately and—in less than frigid weather—run on alcohol alone. On *extremely* cold days, it may be necessary to inject additional gasoline as the engine is running. Since the choke flap is normally pulled partially open by the vacuum draw of the engine, it can be left in a closed position till the engine warms and not interfere with the necessary flow of gasoline.

If your vehicle is equipped with an automatic choke, you can either install a manual conversion kit (available inexpensively at most auto parts stores) or adjust the choke so it remains partially open even when the engine is cold. In

MAKING THINGS GO WITH ETHANOL

severely cold climates, a slightly larger injection jet orifice might be required, but that's just a minor adjustment that can be made while building the injector.

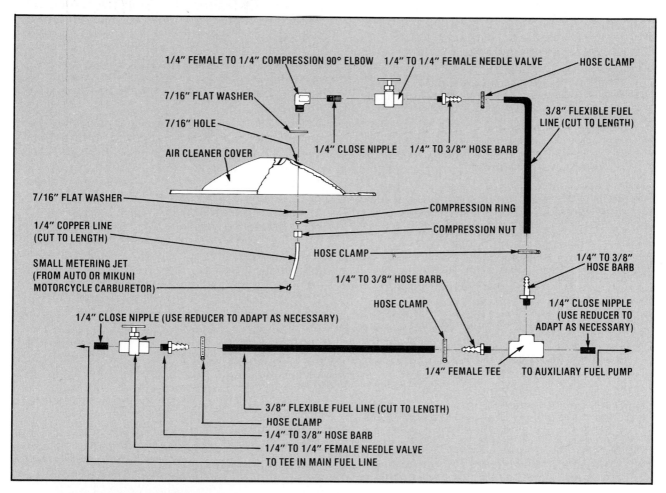

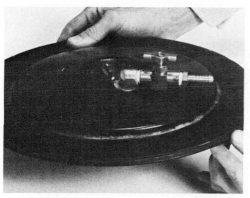

For easier ignition of alcohol fuel in cold weather, you can position a nozzle on the air cleaner lid to inject just the right amount of gasoline into the carburetor.

FUELING AROUND WITH ALCOHOL

Preheating Fuel to Save Time and Energy

While burning alcohol to power a vehicle has the advantage of economy, it does have its drawbacks, especially in cold weather. Because ethanol has a higher vapor point than gasoline, an engine that might run smoothly on the conventional fuel can become rough when fed the less volatile liquid. However, with an uncomplicated, homemade preheater installed in the fuel line, the ethanol will be warmed just prior to entering the carburetor. Thus, vaporization can be increased, smooth operation in frigid weather provided, and fuel economy improved.

Although such a warm-up isn't really necessary in temperate climates, there are folks in parts of the United States—and especially Canada—who will appreciate the benefits that this easy-to-make preheater can provide.

Cold alcohol from the fuel pump—rather than traveling directly to the carburetor—is routed through a length of copper tubing coiled around a short section of copper pipe spliced into the upper radiator hose of an automobile.

As the fluid within the engine's cooling system reaches normal temperature (which may be—depending on the vehicle and the thermostat it has—anywhere from 160° to 195°F), the warm liquid transfers a good deal of its heat to the copper-pipe-and-tubing assembly. The alcohol fuel, then, on its way to the carburetor, picks up the warmth from the coils.

Because ethanol doesn't vaporize as well as does gasoline (and such poor vaporization is aggravated by low temperatures), the preheating device actually serves two purposes: First, it allows warm fuel to enter the carburetor, making that "atomizer's" task easier and preventing cold weather carb problems, and second, the more efficient "mixer" provides increased fuel vaporization within the intake manifold, improving engine economy.

The copper ethanol-warmer takes less than an hour to make and install. Start by cutting a 5" piece from copper pipe that has an external diameter to fit the inside diameter of your upper radiator hose, which must fit snugly around the ends of the pipe. Try to use pipe with an approximately 1/8"-thick wall to prevent the pipe from collapsing when you begin to force the softer tubing around it.

Next, check the diameter of the fuel line, and cut a 42" length of copper tubing with the same diameter (most automobiles use 5/16" or 3/8" line). With steel wool, clean both sections of the tubing you've cut, and wrap the longer length around the 5" piece. Six full coils should be sufficient, but remember to leave a 1-1/2" "tail" at each end of the coiled tube to affix a fuel line inlet and outlet. And allow some room at both ends of the straight pipe to use when you splice your preheater into your cooling line.

Once that's done, heat the 5" pipe with a torch, brush some paste flux onto both copper surfaces, and sweat the coils to the pipe with 50/50 general purpose plumbing solder. Not every loop needs to be completely fastened to the conduit, but it would be a good idea to attach at least the extreme end loops with a continuous bead. The solder not only holds the copper coils fast, but also assists the heat transfer between the tubing and the pipe.

The next step is to install the unit. Locate a 3-1/2" length of rubber radiator hose that fits snugly around the large pipe, clamp the pipe into this flexible coupling, disconnect the existing upper

The flow meter installed on an experimental vehicle that burns alcohol as fuel indicates that fuel consumption is about 4.8 gallons per hour, or 11.4 miles per gallon, 55 MPH.

MAKING THINGS GO WITH ETHANOL

radiator hose at the engine, and fasten the free end of the new hose to the thermostat housing. Complete the junction by connecting the radiator hose to the preheater's free end.

The fuel line can be hooked into the system in much the same way. If your auto already has a flexible neoprene hose, simply cut it and fasten the ends to the coil's inlet and outlet fittings. But if your car's fuel lines are steel, you'll have to remove a section of the line and use two short lengths of neoprene tubing and four clamps to attach the existing line to the warming loops.

As the ethanol fuel passes through the coils, it will naturally gain heat, but even if the engine coolant approaches the boiling point, the warmth will never completely transfer into the moving ethanol. Ideally, the alcohol's temperature should not be allowed to rise above its own boiling point of about 173°F, since this would cause a vapor lock condition within the fuel line and the carburetor. In practice, with six coils incorporated into the preheater, the temperature of the fuel will rise to only about half the temperature of the engine coolant.

As an experiment, the preheating device was installed on a pickup truck, improving its fuel economy by about 11% and raising the average alcohol miles-per-gallon from 10.1 to as high as 11.4 at 55 miles-per-hour.

There's a chance that *your* vehicle's alcohol mileage might be increased further by increasing the number of coils in the preheater, but after a certain point, such additions become ineffective and could result in engine hesitation and power loss when the accelerator is depressed. It may be necessary to experiment until the best balance of engine economy and performance is found.

To make an alcohol preheater, wrap soft copper tubing tightly around a 5″ piece of copper pipe and solder a continuous bead at both ends. You'll also need a 3-1/2″ piece of radiator hose with two correspondingly sized clamps.

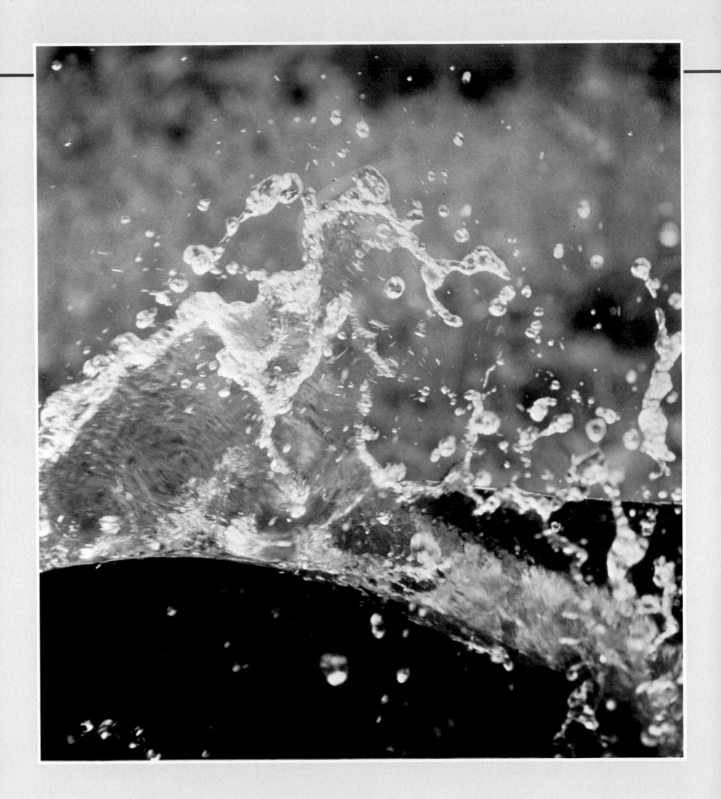

MAKING WATER WORK

Water in motion is potential power . . . a force waiting to be channeled for the accomplishment of any number of mechanical chores. Waterwheels can harness the energy of a stream, using it to power pumps or other machines, or to drive generators that produce electricity. With a water ram, water can be used to pump itself uphill, and with a trompe, the liquid can be used to compress air for use in driving machines in a shop or home.

Homemade power systems are attractive to people with watercourses on their property, but whether such a setup is practical on an individual homestead depends on the work that will be demanded from a waterwheel, whether or not the stream has enough power to accomplish what is desired, and other factors. Fortunately, the answers to most questions can be determined by the landowner before a waterpower project is undertaken.

MAKING WATER WORK

Will Your Stream Work for You?

If you have a stream running across your property, there's a good chance that you've wondered whether you could be generating electricity from the water and whether that homemade power could be saving you dollars. Highly accurate determination of the amount of power available from a given stream is needed to decide on the type of equipment to use. The calculations are relatively complicated and might best be done by a professional, but an individual can make a useful preliminary assessment of a site's potential by using a few basic tools and some elementary mathematics. To determine your hydropower possibilities, just obtain the appropriate information and measurements as outlined in the following steps, make the suggested calculations, and you'll have a fair idea of what your stream can do for you.

Before making any measurements, make certain you are entitled to use your water. In general, there are two types of water rights used in the United States: riparian and appropriative. The former is commonly employed in eastern states and stipulates that, unless otherwise noted, water rights are transferred with the deed to a piece of property. Consequently, if you live east of the Mississippi River, you probably have the basic right to use the water on your land. On the other hand, appropriative rights are used in a number of western states (Alaska, Arizona, Colorado, Idaho, Montana, Nevada, New Mexico, and Utah) and in combination with riparian rights in eight other states (California, Kansas, Nebraska, North and South Dakota, Oregon, Texas, and Washington). Your right to use water in states which have opted for appropriative rights is far from certain—in some cases water privileges remain with the property's first owner, while in others, rights can be lost from nonuse—so you must check carefully to ascertain if the water is yours to use.

In addition, you should get in touch with the local office of the Army Corps of Engineers to determine if your stream has been declared navigable and therefore falls under Corps jurisdiction (you could be surprised). In such a case, your right to tap a Corps of Engineers stream is dependent upon your ability to convince them that you won't alter the flow of the river. However, with the recent encouragement of small hydropower by the Department of Energy, you're likely to find the Corps cooperative and helpful. They have expertise to offer in civil works, and they can supply flow records, drainage patterns, and rainfall figures for your site.

Of the two basic components which make up the energy of waterpower—flow and head—flow is by far the more difficult to measure accurately.

Small flows can be measured by capturing the moving water in a barrel. For this you'll need a way to route all the water into a vessel, and you'll have to record the length of time it takes for the receptacle to fill. For example, if a 55-gallon drum fills up in 60 seconds, the flow is 55 gallons per minute. However, when working with hydropower, we usually consider flow in cubic feet rather than gallons, so 55 gallons per minute is translated to cubic feet per minute by dividing it by the number of gallons in a cubic foot—7.5—to give 7.33 cubic feet per minute.

If your stream is so small that you can actually use the container method, you must have very high head to get any useful amount of power. In the example above, theoretically, at least 100 feet of head would be needed to produce one kilowatt of electricity.

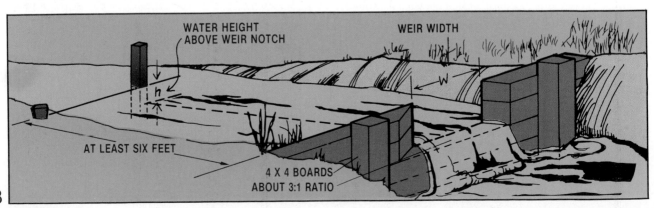

WATER HEIGHT ABOVE WEIR NOTCH — WEIR WIDTH — h — W — AT LEAST SIX FEET — 4 X 4 BOARDS ABOUT 3:1 RATIO

HYDROPOWER

The flow of a stream is a function of both its cross-sectional area *and* the velocity of its water. Hence the flow can be determined by measuring depths at one-foot intervals and adding them together, multiplying that figure by the width, and then multiplying the result by the water's speed as measured by a float moving over a 30-foot section of stream.

Unfortunately, the inaccuracy of measuring the area in this fashion when coupled with the errors introduced by an irregular stream bed, which produces what is called nonlaminar flow, may result in a float method flow calculation that is off by as much as 40%. Consequently, the technique is generally reserved for making a basic judgment on whether there's enough flow to make power generation possible, and the method is not suitable for determining equipment size. However, since we're only interested in getting a *basic* idea of the stream's capacity here—and because the float method is so easy to use—it's quite well suited to our present purpose. Plug your numbers into the following formula for a guiding hand:

_____ square feet (cross-sectional area) × _____ feet per second (velocity) = _____ cubic feet per second (flow) × 60 = _____ cubic feet per minute (flow)

The most accurate approach to gauging flows that are too large for a container is to build a weir. This involves constructing a dam across the stream, with a spillway that's large enough to accommodate all the flow without it spilling over the water barrier's top. Obviously, building such a dam is a relatively involved process, so you may want to use it only after you've demonstrated that your stream has solid potential for hydropower by the float method. (Certain turbine types require such an accurate measurement, while for others a person can get by with the approximate figure derived by the float method.)

The theory behind a weir is that the amount of water which backs up behind the barrier (measured in inches of depth above the notch in the dam) is representative of the flow. After you've built the weir, you'll have to drive a stake into the stream's bottom (about six feet upstream) and measure the difference in height between the weir notch and the water level on the stick. Then apply that number to the Weir Table. Take the flow figure supplied by the table and multiply it by the width of the weir notch in inches to arrive at an accurate flow rate for your stream.

WEIR TABLE			
DEPTH ON STAKE IN INCHES	CU. FT. PER MIN. PER INCH LENGTH	DEPTH ON STAKE IN INCHES	CU. FT. PER MIN. PER INCH LENGTH
1	0.4	10	12.7
1.5	0.7	10.5	13.7
2	1.1	11	14.6
2.5	1.6	11.5	15.6
3	2.1	12	16.7
3.5	2.6	12.5	17.7
4	3.2	13	18.8
4.5	3.8	13.5	19.9
5	4.5	14	21.1
5.5	5.2	14.5	22.1
6	5.9	15	23.3
6.5	6.6	15.5	24.5
7	7.4	16	25.7
7.5	8.2	16.5	26.9
8	9.1	17	28.1
8.5	10.0	17.5	29.4
9	10.8	18	30.6
9.5	11.7	18.5	31.9

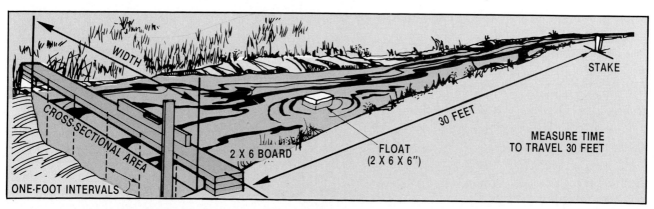

WIDTH

CROSS-SECTIONAL AREA

ONE-FOOT INTERVALS

2 X 6 BOARD

FLOAT (2 X 6 X 6")

30 FEET

STAKE

MEASURE TIME TO TRAVEL 30 FEET

MAKING WATER WORK

Will Your Stream Work for You? (continued)

Once the rate of flow for a stream is established, it's necessary to know the total vertical distance (in feet) the water must travel to reach the centerline of a turbine or other water-driven device. This is known as the nominal, or gross, head. Some types of turbines *can* be encased in

HEAD LOSS IN FEET (PER 100 FEET OF PLASTIC PIPE[1])								
FLOW (cubic feet per minute)	PIPE DIAMETER (INCHES)							
	2″	2.5″	3″	4″	6″	8″	10″	12″
3	1.8	0.6	0.2	0	0	0	0	0
6	6.3	2.1	0.9	0.2	0	0	0	0
12	23.0	7.5	3.0	0.7	0.1	0	0	0
18	•	16.1	6.4	1.5	0.2	0	0	0
24	•	27.4	11.0	2.6	0.4	0	0	0
30	•	•	16.4	4.0	0.5	0.1	0	0
36	•	•	23.4	5.6	0.7	0.2	0	0
42	•	•	31.2	7.4	1.0	0.2	0.1	0
48	•	•	•	9.5	1.3	0.3	0.1	0
54	•	•	•	11.9	1.6	0.4	0.1	0
60	•	•	•	14.4	1.9	0.5	0.2	0.1
66	•	•	•	17.2	2.3	0.5	0.2	0.1
72	•	•	•	20.1	2.7	0.6	0.2	0.1
78	•	•	•	23.0	3.0	0.7	0.2	0.1
84	•	•	•	26.8	3.5	0.9	0.3	0.1
90	•	•	•	30.5	4.0	1.0	0.3	0.1
120	•	•	•	•	6.9	1.6	0.6	0.1
150	•	•	•	•	10.5	2.4	0.8	0.3
180	•	•	•	•	15.5	3.4	1.4	0.5
240	•	•	•	•	25.8	6.1	2.0	0.8

[1] Double for steel or black iron.
• Losses exceed gains.

the pipe between intake and exhaust, thereby benefiting from what's called "suction" head. When you calculate head, be sure to make allowances for the mounting of the equipment and for variations in water level.

You may want to employ a surveyor to get a very accurate measurement of head, or you can do a fairly good job yourself by using one of a couple of "seat of the pants" approaches. If your site's head will exceed 15 feet, you can get adequate numbers by using a homemade transit consisting of a tripod and a hand level. Or as an alternative you can use the old foundation leveling garden hose method where one end of the water-filled tube is dragged uphill and the other end is lifted to keep both ends at the same height. Since the water seeks its own level, the height of the hoisted end can be measured and will be equal to the height of the end strung up the hillside.

Unfortunately, the friction of the pipe used to channel the water to a turbine results in some reduction in velocity—or head loss—and, accordingly, reduced energy at the generator. Hydrotechnical engineers have formulas for calculating such losses, but it's most convenient for us to refer to a head loss table which provides us with energy loss figures in feet of head. Head loss is merely subtracted from nominal head to yield the net head:

_____ feet (gross head) − [_____ feet (head loss per 100 feet) × _____ hundreds of feet] = _____ (net head)

Now you can begin to see the potential of your stream. The electrical power capacity of a hydropower plant is determined by the amount of horsepower that falling water can develop at the turbine

HYDROPOWER

(1 HP = 33,000 foot-pounds per minute = .746 kw). Of course, in hydropower calculations we've transformed the foot-pounds into cubic feet of H_2O flow per minute, so the formula looks slightly different:

$$\frac{flow \times head}{709} = kw\ potential$$

So, with your flow and net head already determined, just fill in the numbers:

_____ cubic feet per minute (flow) × _____ feet (head) ÷ 709 = _____ kw potential

However, kw potential is a measurement of how much power is available to spin the turbine, and some of that power will be lost by the time we generate electricity with it. Specifically, the losses at the turbine, in gearing, and from the generator/alternator will be a minimum of 20%, and to be conservative you should figure your system to be about 75% efficient. So:

_____ kw potential × .75 efficiency = _____ kw net

Now, to figure the value of this potential power, look up one of your recent bills and note the rate that you pay for electricity per kwh. (Your utility may not specify, because the rate varies so much, but an average figure will do fine.) Then multiply the price you would have to pay per kwh times the number of kwh that a hydroplant can supply you per year. (The modern hydroelectric system is extremely reliable and can be expected to perform 24 hours per day for 350 days of the year . . . with only about a day per month down-time for intake cleaning and lubrication.) The calculation for this step is as follows:

$_____ kwh (price) × _____ kw capacity per hour × 24 hours × 350 days = $_____ per year's worth of kwh

Perhaps the main thing on your mind is the amount of money you can save installing hydroelectric equipment. The answer is: nothing . . . for a while. Normally the economics of hydropower are determined by calculating the length of time it will take for the potential energy savings to pay off the capital you'll invest in the system. An easier way to figure return on investment is to decide what you think is a good payback period and then compute from that how much you can pay for your water power system.

For demonstration purposes, let's choose a seven year return on investment, since that's an average payback nowadays. In order to determine how much money you can spend for a given amount of power and still get your cash back in seven years, take the annual production in dollars' worth of kwh, add about 7% inflationary increase per year, and total the payoff for those seven years. For instance, a unit that produces $875 worth of power this year will crank out a grand total of $7,567 worth of watts over the next seven years, or enough to purchase a system capable of supplying an average household.

MAKING WATER WORK

Stream and Spring

By tapping a spring for drinking and a stream for power, people can have their water and pump it too. One successful example of how this can be done is on a farm near Halifax, Virginia where a 120-foot-long flume carries water from a mill-pond to turn a 6-1/2-foot-diameter waterwheel which, in turn, drives a pump that delivers 1,440 gallons of pure springwater 100 feet uphill to the tank of a gravity-fed water system that serves the family homestead.

Unlike a hydraulic ram, this system doesn't use a great quantity of drinking water to pump just a few gallons uphill. The overshot wheel makes efficient use of the small stream that's been tapped for power and leaves the fresh water supply unaffected. Moreover, the setup requires very little maintenance. Aside from the initial investment, the only expense is for a few shots of grease for the bearings about three times a year.

Construction of the homebuilt system was begun by building a dam about three feet high across a stream that ran through the property. Since about seven feet of fall is needed to turn the wheel, water from behind the dam was piped to a spot 120 feet downstream where the creek's banks are four feet lower. This four feet plus the dam's three feet provides the required fall or head.

The flume consists of approximately 85 feet of 4'' aluminum pipe, which empties into a wooden trough six inches wide, six inches deep, and 46-1/2 feet long. The aluminum pipe extends about a foot and a half into the wooden trough and can be lifted out and set into a curved metal deflector which routes the water back into the stream whenever the owners want to stop the wheel. Though traditional homestead flumes are of all-wood construction, corrosion-free aluminum pipe was chosen for

the first two-thirds of the wheel's feeder line because parts of this section of the flume run underground, and the upper end extends into the water. The pipe's inlet was covered with a screen to keep the line from becoming clogged with errant leaves, turtles, and crawfish. The lower end of the flume was carefully positioned so that the discharged stream of water would hit the exact top-center of the overshot wheel.

The parts of the wheel were prefabbed by a steel company and then assembled by the owner. Each of the 980-pound assembly's 37 buckets consists of 3/16'' steel plate cut 8'' X 12''. The hubs were cast at a local foundry from a wooden pattern made by the owner, and the axle holes were bored at a machine shop. The 5''-wide rims and the 12''-wide inner plate were also fashioned from 3/16'' steel, and each of the 12 spokes (six on each side) is steel, 3/8'' thick and 2'' wide.

Once all the components were ready, the wheel was assembled. The buckets were very carefully spaced and tacked into place before they—and the back plate—were welded to the rims. Then the spokes were fastened between the rims and the hubs. The radials were welded at the outer edge and bolted to center castings. The 16-inch distance between the hubs is four inches greater than the wheel's 12-inch width at the rim, creating a dish effect which adds a great deal of strength to the apparatus.

The completed waterwheel was lifted into place on poured concrete foundations, where its 2''-diameter shaft was secured in place at both ends in pillow-block bearings. A few small pieces of steel were welded to some of the buckets to balance the assembly, and a coat of paint gave the wheel some rustproofing.

HYDROPOWER

The wheel was located beside a reservoir that holds springwater piped from its source about 120 feet away and slightly uphill. This masonry box contains the pump, which lifts the potable water, pushing it 100 feet uphill to a 500-gallon storage tank that sits on a tower outside the family residence. From the tank, water flows into the homestead plumbing system, all of which is lower than the water tower and receives the water through gravity flow from the tank.

To lift the water from the reservoir to the tank, it was necessary to tie the waterwheel to the pump. To do this, a 15″ gear salvaged from a local junkyard was mounted on the wheel's axle. A hole was drilled in the outside face of the gear, 1-1/2 inches from the center, and it was tapped to accept the threaded end of a steel mounting pin. As the wheel turns, the pin turns in a three-inch circle, raising and lowering a long, white oak connecting rod whose lower end is mounted on the pin. The upper end of the connecting rod is attached to one end of a horizontal arm, the other end of which is secured to a red cedar post so the arm can swing up and down as the connecting rod to the wheel rises and falls. A cylinder rod is attached just off-center to the middle of the cross arm and runs down to drive the pump, which is anchored to the bottom of the reservoir.

The continuous flow of the 52°F springwater prevents it from freezing, and the system provides an adequate supply of water not only to the house, but also to the farm's stables and swimming pool.

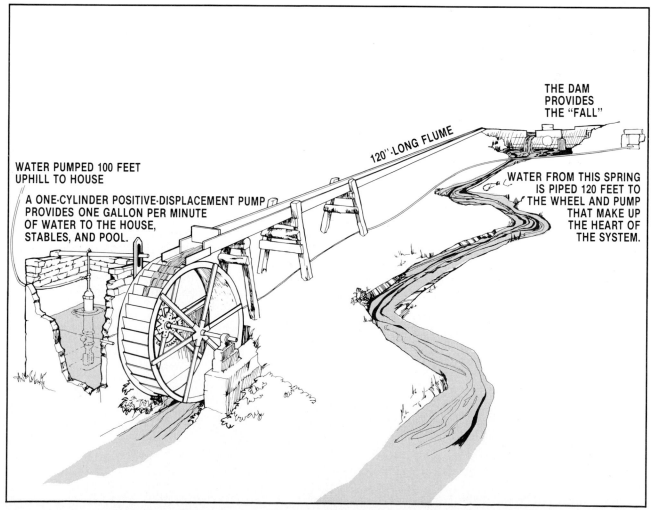

WATER PUMPED 100 FEET UPHILL TO HOUSE

A ONE-CYLINDER POSITIVE-DISPLACEMENT PUMP PROVIDES ONE GALLON PER MINUTE OF WATER TO THE HOUSE, STABLES, AND POOL.

120″-LONG FLUME

THE DAM PROVIDES THE "FALL"

WATER FROM THIS SPRING IS PIPED 120 FEET TO THE WHEEL AND PUMP THAT MAKE UP THE HEART OF THE SYSTEM.

MAKING WATER WORK

Home Hydropower
Electricity by the Gallon

Despite the fact that water-generated power accounts for a mere 11% of the electricity consumed in the United States, hydropower is one of the least expensive (and, perhaps, most environmentally benign) sources of electricity available in the nation. And in fact, though there seems to be a mounting resurgence of interest in hydroelectric energy in this country (30 years ago it supplied 15% of our needs), it is *not* coming from huge dams and multimegawatt-producing turbines, but is happening on a small—and in many cases, private—scale.

The combination of ever-increasing kilowatt-hour costs, electronic improvements that allow the construction of automatic hydropower controls, and available commercial tax credits for renewable energy is making it economically attractive for landowners with adequate creeks to develop their own power-producing sites.

Once you've determined that it will be economically feasible to use your water course for a hydroelectric plant (see page 88), you may find it helpful to evaluate the problems and challenges faced by the owners of one particular site. This property already contained a dam and a reservoir that were deemed very suitable for the installation of a hydropower system, so the first item of business was to search for the equipment needed to build a plant.

There were a number of factors that limited the potential sources of equipment. For one thing, because the dam was only 15 feet high, most of the standard packages, which include high-head Pelton wheels, were out of the question. Furthermore, the nearest structure for which the waterwatts could be used was already built and wired for standard AC voltage. So a decision to be independent of any utility demanded that the prospective home energy producers employ either a DC generator and an inverter, or an alternator.

The cost of both these approaches proved to be far beyond the funds budgeted for the project, so at that point, a hydropower consulting firm was asked to provide assistance. The task of the company's hydrotechnical engineer was to come up with an acceptable system in terms of both capability and price, and his solution was a combination of a number of established concepts that formed an exceptional package. For the many people faced with similar problems, the ideas he presented could represent a bona fide home hydropower breakthrough.

One of the most expensive components in any hydroelectric plant is the sophisticated rotor that extracts power from moving water. Though the cost of casting or joining the pieces for a turbine varies with the type, the runner generally absorbs between one-third and one-half of a small installation's entire budget. To get around this expense, the engineer combined the knowledge gained during his years of experience working with crossflow turbines in Europe with a body of information on build-it-yourself turbines assembled by the engineering department of Oregon State College and Volunteers in Technical Assistance (VITA). The result is a Mitchell-Banki–style turbine that was built from scrap 4"-nominal mild steel pipe, 1/4" steel plate, and a 1-7/16" steel shaft.

The runner itself is 12" in diameter and 18" long, and has 20 blades (formed from 72° arcs of the 4" pipe). The unit's shaft rides on a pair of standard pillow-block bearings, and water is delivered through a nozzle which can be altered to accommodate a wide range of flow rates.

HYDROPOWER

After spinning the turbine, which extracts some of the stream's energy, the water that drives the powerplant cascades downward and is returned to the streambed. The entire process provides electricity, yet leaves the environment virtually undisturbed.

MAKING WATER WORK

Home Hydropower
Electricity by the Gallon (continued)

The turbine spins at 275 RPM and is connected to the alternator by three-groove pulleys and V-belts, which increase the RPM to 1,800. Though that rotational speed is perfectly suited to electrical generation, the AC system requires precise RPM control in order to maintain consistent power cycling. And, since changing loads and flow variations *do* tend to affect turbine speed, it initially seemed necessary to compromise power output (by spilling some of the flow or employing a clutch most of the time) in order to compensate for speed and load changes.

In order to avoid this compromise, an electronic device called a load-ballast speed control was installed. The circuitry in this device accomplished two tasks: First, it maintains as much output as the alternator can handle without going above or below the desired speed (1,800 RPM). That power output is directed either to a water heater (the ballast) or to the circuit where power is demanded. A solid-state electronic switch called a triac then juggles the power, in minute increments, between the ballast and demand circuits to maintain constant output.

The system's power-producing package is an off-the-shelf self-exciting alternator. It is designed to deal with loads up to 2.5 kw on a continuous basis and provides one 120-volt, 60-cycle circuit.

Depending upon the rate of inflation, a system of this design should be able to pay off the initial investment in under ten years. After that point the water-produced electricity will be practically free until the system is worn out.

But perhaps the most enjoyable aspect of the installation is the thought that the electric power it's generating comes from the natural movement of water downhill. Nothing was burned, and no atoms were split to make the power. It's just capturing the energy that otherwise would have kept on flowing along.

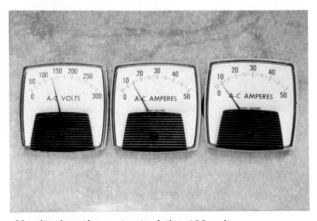

Monitoring the output of the 120-volt, 60-cycle alternator is done through gauges that indicate the voltage being produced and the current being used by the system at any given time.

THE MANY USES OF WATER

Homemade Hydraulics
The Water Ram Pump

The hydraulic ram has been around for quite a while and was a widely used means of pumping water before electricity became common in rural areas. Strictly speaking, this type of pump doesn't create its own power, but draws energy from the force of a moving column of water, usually fed through a pipe from a point more than 18 inches above the ram.

The diagram (on this page) illustrates the simplicity of the water ram's operation. Water from the feed pool or reservoir rushes down the drive pipe—flowing past the waste valve and out the waste pipe—until enough pressure builds up to force the bathtub stopper against its seat. (Naturally, this pressure increases as the fall from the source becomes greater.)

When the waste valve shuts, it drives water through the check valve and into an air chamber, where the fluid compresses the air and forces it to kick back, like a piston. This action, in turn, closes the check valve and pumps water out the delivery pipe and—eventually—into a pond, tank, or irrigation system.

As the check valve closes, the water in the drive pipe rebounds temporarily, creating a partial vacuum that allows the waste valve to drop open again. The excess liquid (which was not driven into the delivery pipe) then flows out the waste opening and can be returned to the water source or used to fill another pond. (The point of return must be lower than the intake point.)

The entire cycle is repeated some 40 or more times per minute. The rate can be increased by tightening the inner adjusting nuts, while a greater flow of water will be achieved by *loosening* the adjusters. Of course, you'll have to "fine tune" your ram to suit your particular needs.

Although the pump won't operate without a fluid power source, it *will* work indefinitely when water is present. And, except for an adjustment every few months, the device requires no maintenance whatsoever.

Quality hydraulic rams are available commercially for anywhere from $400 to $800, and these are excellent units designed to last for many

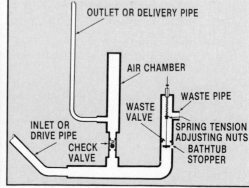

The water ram pump will work indefinitely when water is present, and except for an occasional adjustment, the device requires no maintenance.

OUTLET OR DELIVERY PIPE

AIR CHAMBER

WASTE PIPE

WASTE VALVE

SPRING TENSION ADJUSTING NUTS

INLET OR DRIVE PIPE

CHECK VALVE

BATHTUB STOPPER

MAKING WATER WORK

**Homemade Hydraulics
The Water Ram Pump (continued)**

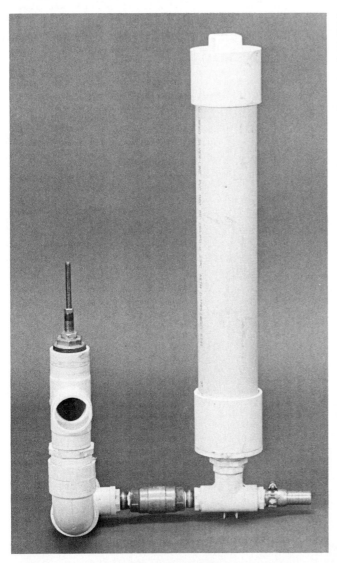

years. But since most folks don't have that kind of money to spend, the homebuilt pump shown here was designed to deliver the same reliable performance as one of its store-bought cousins, but at only a fraction of the cost. Better yet, it can be put together in less than an hour and is made of off-the-shelf plumbing supplies.

Most of the components of this pump are fabricated from PVC pipe . . . the same conduit that's used throughout the country in residential and industrial plumbing systems. Polyvinyl chloride is tough, easy to work with, and readily available. The moving parts of the ram consist of a standard check valve, a steel rod, a spring, and a rubber bathtub stopper.

Begin assembly of your ram pump by purchasing or otherwise obtaining all of the plumbing components that are detailed in our materials list (don't forget to buy a can of PVC cement to permanently seal all the joints in the system). Then find a bolt that measures 1/4" X 14", cut the head off, and thread the "new" upper shank to a depth of about three inches. Threaded rod won't do here, since the shaft must have a 4" (at least) smooth-surface midsection.

Next, run two hex nuts from the lower end of the rod to a point about 5 inches up the shaft, and fasten the bathtub plug to the tip of the stem with nuts and washers as illustrated. With this done, thread the "waste" portion of the pump together (the parts from the 1/4" X 1-1/2" brass nipple down to the 2" PVC to 2" pipe bushing on the assembly drawing). Then install the washer and spring on the upper part of the 14" rod, and slip this entire waste-valve assembly up into its PVC housing and *through* the brass nipple at the top. Lock the completed valve in place with the remaining 1/4" nut.

LIST OF MATERIALS

3" PVC threaded plug	1-1/2" pipe to 1-1/2" hose galvanized adapter
3" pipe (female) to 3" PVC (female) coupling	1-1/2" pipe (female) to 1-1/2" (male) bushings
3" PVC pipe (19-1/2" long)	1-1/2" PVC (female) to 2" PVC (male) bushing
3" PVC coupling	1" X 2" X 2" PVC tee
3" PVC (male) to 1-1/2" PVC (female) bushing	2" PVC street elbow
1-1/2" PVC (male) to 1-1/4" PVC (female) bushing	2" PVC (male) to 2" pipe (female) bushing
1-1/4" PVC (male) to 1" PVC (male) nipple	2" pipe (male) to 1-1/2" pipe (female) bushing
3/4" pipe to 3/4" hose galvanized adapter	1-1/2" pipe (male) to 1-1/2" PVC (male) nipple
3/4" pipe (female to 1" PVC (male) bushings	1-1/2" X 1-1/2" X 1-1/2" PVC tee
1" X 1" X 1" PVC tee	1/2" pipe (female) to 1-1/2" pipe
3/4" X 1-1/2" nipples	(male) galvanized bushing

THE MANY USES OF WATER

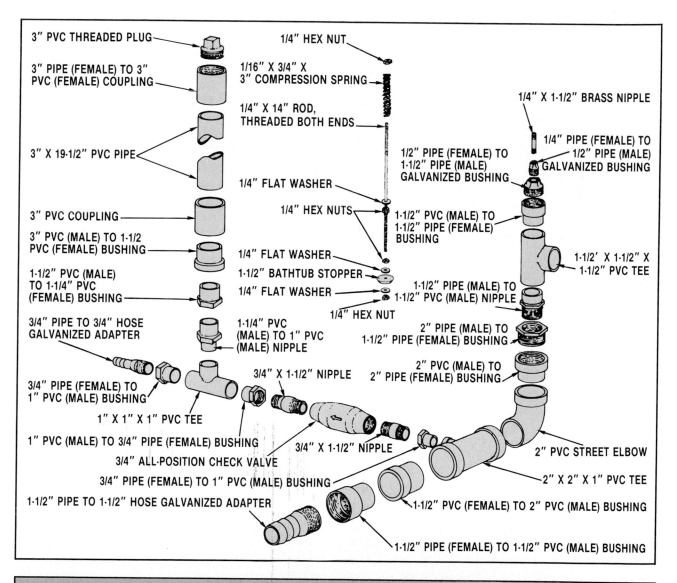

3" PVC THREADED PLUG

3" PIPE (FEMALE) TO 3" PVC (FEMALE) COUPLING

3" X 19-1/2" PVC PIPE

3" PVC COUPLING

3" PVC (MALE) TO 1-1/2 PVC (FEMALE) BUSHING

1-1/2" PVC (MALE) TO 1-1/4 PVC (FEMALE) BUSHING

3/4" PIPE TO 3/4" HOSE GALVANIZED ADAPTER

3/4" PIPE (FEMALE) TO 1" PVC (MALE) BUSHING

1" X 1" X 1" PVC TEE

1" PVC (MALE) TO 3/4" PIPE (FEMALE) BUSHING

3/4" ALL-POSITION CHECK VALVE

3/4" PIPE (FEMALE) TO 1" PVC (MALE) BUSHING

1-1/2" PIPE TO 1-1/2" HOSE GALVANIZED ADAPTER

1/4" HEX NUT

1/16" X 3/4" X 3" COMPRESSION SPRING

1/4" X 14" ROD, THREADED BOTH ENDS

1/4" FLAT WASHER

1/4" HEX NUTS

1/4" FLAT WASHER

1-1/2" BATHTUB STOPPER

1/4" FLAT WASHER

1/4" HEX NUT

1-1/4" PVC (MALE) TO 1" PVC (MALE) NIPPLE

3/4" X 1-1/2" NIPPLE

3/4" X 1-1/2" NIPPLE

1/2" PIPE (FEMALE) TO 1-1/2" PIPE (MALE) GALVANIZED BUSHING

1-1/2" PVC (MALE) TO 1-1/2" PIPE (FEMALE) BUSHING

1-1/2" PIPE (MALE) TO 1-1/2" PVC (MALE) NIPPLE

2" PIPE (MALE) TO 1-1/2" PIPE (FEMALE) BUSHING

2" PVC (MALE) TO 2" PIPE (FEMALE) BUSHING

1/4" X 1-1/2" BRASS NIPPLE

1/4" PIPE (FEMALE) TO 1/2" PIPE (MALE) GALVANIZED BUSHING

1-1/2' X 1-1/2" X 1-1/2" PVC TEE

2" PVC STREET ELBOW

2" X 2" X 1" PVC TEE

1-1/2" PVC (FEMALE) TO 2" PVC (MALE) BUSHING

1-1/2" PIPE (FEMALE) TO 1-1/2" PVC (MALE) BUSHING

LIST OF MATERIALS (continued)

1/4" pipe (female) to 1/2" pipe (male) galvanized bushing
1/4" X 1-1/2" brass nipple
1/4" X 14" bolt
1/4" hex nuts
1/4" flat washers
1/16" X 3/4" X 3" compression spring
1-1/2" bathtub stopper
1/2 pint PVC cement
3/4" flexible polyethylene pipe
1-1/2" flexible polyethylene pipe

close-knit protective filter screen
hose clamps
3/4 all-position check valve

NOTE: If you wish, you may substitute Schedule 40 black pipe (of comparable dimensions) for the flexible polyethylene used at the drive and delivery ends of the systems.

MAKING WATER WORK

Homemade Hydraulics
The Water Ram Pump (continued)

With this section completed, merely assemble the rest of the plumbing, using our illustration on page 99 as a guide (remember that the PVC joints must be glued together if they're not already threaded). Then you can either let the ram stand by itself, or—for a more permanent fixture—mount it to a block of wood by using two lengths of hanger strap fastened with wood screws.

The hydraulic ram is as easy to set up as it is to build. It requires only that the stream, pool, or other water source be at least 18 inches *above* the pump and that it will provide a flow of no less than three gallons per minute into the ram.

When you've determined the water supply to be adequate, install the ram at a point *no less* than a foot and a half below the source (the higher the source, the greater the output will be). The length of 1-1/2" tubing from the supply to the pump, known as the drive pipe, should be from 10 to 15 times greater than the distance of fall, and—to avoid drawing foreign matter into the machinery —a filter screen must be placed over the drive pipe's inlet opening.

Now, simply run the necessary length of 3/4" flexible polyethylene tubing (the durable, high-pressure type) to your storage tank or reservoir from the pump outlet (this is the delivery pipe). Take care not to allow any kinks or sharp bends to form in the hose, since such "corners" will impede or stop the flow of water. You can also route a 1-1/2" line from the ram's "waste" pipe back into the feed stream, or—if you wish—you can divert this water for some other purpose.

You'll be amazed at how well the simple pump performs. The test ram, with an 11.5-foot fall, makes a steady supply of water available at a point 65 feet above the hydraulic mechanism. And with a 40-foot fall, a whopping 60 gallons per hour can be delivered from the outlet conduit: enough to supply a reservoir with a total of 1,440 gallons in one 24-hour period.

Because of the friction factor, the total length of the delivery pipe has a bearing on the performance of the pump. As a rule, this tube should be no more than 20 times the height that the fluid is to be lifted. In most situations, this "formula" is easy to live with.

Naturally, since the conditions under which each pump operates will vary, the mechanism will have to be adjusted to suit individual needs. Do this by simply putting an end wrench through the ram's waste pipe opening and turning the inner hex nuts to either tighten or loosen the tension on the spring. More "pressure" will increase the rate of waste-valve "action" (simultaneously *decreasing* the amount of water pumped), while *relieving* the spring tension will force *more* fluid through the check valve—and out the ram's delivery pipe—with each stroke.

When you stop to think about it, it's a wonder that the hydraulic ram isn't being used nearly everywhere. After all, the design is virtually maintenance-free, creates no waste or pollution, and costs next to nothing to build and install. Furthermore, by the use of larger or smaller plumbing components, the hydraulic pump can be scaled up or down to suit individual needs.

So—if you want to supply a farm pond with fresh water, fill a storage tank, deliver drinking water, or feed an irrigation system—the simple water ram can do the job . . . and *without* the outside power requirements and maintenance headaches that are unavoidable when using conventional pumps.

THE MANY USES OF WATER

An Ace of Trompes, the All-purpose Water Machine

The *trompe*, which dates back to the beginning of the Iron Age, is a device that uses the energy of falling water to pressurize air. This is achieved by means of a standpipe or shaft down which a column of water is allowed or directed to fall. As it drops, the water draws air through small inclined orifices and carries it to a submerged plenum or reservoir, where the air separates from the water and is held under pressure. The water—meanwhile—continues to flow to an exit pipe whose end is high enough to balance the pressure in the reservoir. The pressurized air can then be drawn off through a tuyere—or escape nozzle—to be used as needed.

For the homesteader or farmer with a good-sized stream on his or her property, the trompe offers a virtually inexhaustible supply of free, cool, compressed air that can be used to operate a forge, to drive machinery, or to air-condition a house or barn. Except for a simple water-flow control, trompes have no moving parts, rely on no computerized technology, make no noise, and they don't pollute the environment.

One remarkable feature of trompes is that the air that comes out of the system is actually cooler than the air that enters it. This is because the cold water flowing through the trompe absorbs the heat that's usually generated by the compression of air. Therefore you get air that's the same temperature as the cool water it just left, and the result can be free air-conditioning.

There are a couple of things to remember if you decide to construct your own hydraulic air compressor:

[1] If you want to be able to switch the air off but leave water running through the trompe, build a blow-off pipe and locate the lower end in the plenum slightly below the normal water level. Then when the airflow through the discharge line is blocked, the air pressure in the reservoir will increase to the point at which the water level drops and surplus air can escape through the blow-off pipe.

[2] When you first start up your trompe, keep the air from escaping through the discharge line until the pressure has had a chance to build. Otherwise you'll end up with a plenum full of water and no compressed air.

The trompe is an old but good idea, and its usefulness knows few limits.

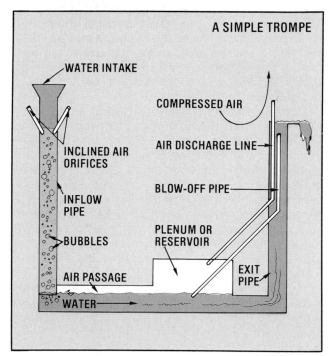

A SIMPLE TROMPE

WATER INTAKE — INCLINED AIR ORIFICES — INFLOW PIPE — BUBBLES — AIR PASSAGE — WATER — COMPRESSED AIR — AIR DISCHARGE LINE — BLOW-OFF PIPE — PLENUM OR RESERVOIR — EXIT PIPE

MAKING WATER WORK

A Moving Experience
The Home-sized Waterwheel

What if you don't need compressed air or hydroelectricity, but you do need a reliable supply of water? If you live near a stream, you can obtain access and control of that precious liquid with a waterwheel.

A miniature adaptation of the familiar paddle wheel can deliver gallons of "Adam's Ale" to your homesite in a single day. This 48"-diameter device—made out of 3/4" marine plywood and pressure-treated 1 X 6 board—drives a recycled shallow-well pump which, in turn, forces water up a hill to a 1,000-gallon cistern. The mini-wheel is small but effective: It pushes one gallon of water a minute into the underground holding tank. At that rate the setup delivers more than 1,000 gallons daily (far more than most families use), so there's always plenty of water in the cistern. And if the storage tank is located several feet above the site where the water's used, there's always plenty of pressure.

The key to this system is that it uses water from one source to pump water from another source. The little stream that powers this particular waterwheel is an open waterway, and the local building codes wouldn't allow anyone to draw drinking liquid from such an easily contaminated source. So potable water had to be obtained from a capped spring 740 feet away. One-inch plastic pipe carries this trickling flow down to the waterwheel's pump, which—driven by the force of the open, "dirty" stream— lifts the clean water to the storage tank.

The waterwheel is built in the classic "overshot" style (that is, a flume shoots water to the top of the wheel, where the fluid falls into built-in buckets forcing the rig to turn). To get the power-providing liquid into "overshot position", a pair of 1-1/2" plastic pipelines were run far enough up the stream to capture water at an altitude slightly above the top of the wheel and then to dump it right at the top of the bucketed double disk (at a rate of about 45 gallons per minute).

The design of the water-pumping system is pretty straightforward. Its most complex element is the wheel itself, and even that has a symmetrical simplicity. To construct the rotating unit, first cut two 48"-diameter circles out of 3/4" marine-grade plywood. Then scribe pencil lines that divide each "pie" into 16 equal sections. Each point at which one of those lines meets the perimeter of the circle marks the location of the outside edge of one of the wheel's 16 water-catching scoops. A chord drawn between any adjacent pair of these points measures 9-3/8".

The water will exert *maximum* turning force on the wheel if every bucket holds its contents as long as possible. Consequently, the dividing boards need to be laid out so that they angle steeply up toward the flume's spillway. That way, an individual container will travel more than one-third of of a revolution before its lower side becomes horizontal and dumps its contents.

To set the desired angle for each bucket, place a framing square so that one side of the tool lies on one of the 16 radial lines, while its other side intersects the outer tip of the second radial line back on the wheel. The board is placed along the line thus formed, to connect the two radii. Each 1 X 6 section thereby forms a 90° angle with a radial line at its inner end and works out to be 19-1/2" long. In addition, since every board overlaps a neighboring plank, it serves as both the front side of one bucket and the back piece of the next!

THE MANY USES OF WATER

The steeply slanted "buckets" hold
their contents until they've nearly
reached the stream, allowing the water
to exert maximum force on the
small-scale waterwheel.

A Moving Experience
The Home-sized Waterwheel (continued)

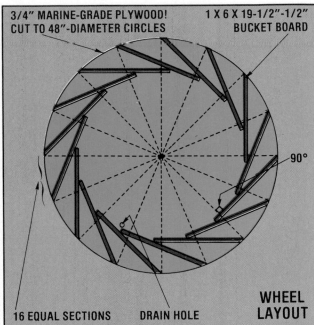

3/4" MARINE-GRADE PLYWOOD! CUT TO 48"-DIAMETER CIRCLES

1 X 6 X 19-1/2"-1/2" BUCKET BOARD

90°

16 EQUAL SECTIONS   DRAIN HOLE

WHEEL LAYOUT

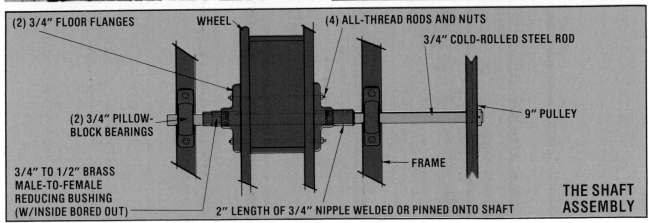

(2) 3/4" FLOOR FLANGES

WHEEL

(4) ALL-THREAD RODS AND NUTS

3/4" COLD-ROLLED STEEL ROD

(2) 3/4" PILLOW-BLOCK BEARINGS

9" PULLEY

3/4" TO 1/2" BRASS MALE-TO-FEMALE REDUCING BUSHING (W/INSIDE BORED OUT)

FRAME

2" LENGTH OF 3/4" NIPPLE WELDED OR PINNED ONTO SHAFT

THE SHAFT ASSEMBLY

To increase the pumping rate of the wheel, the designers put the 9″ pulley that came on their shallow-well pump on the bucket-bearer's shaft and replaced the 9″ pulley with a 2-1/2″ disk.

THE MANY USES OF WATER

After all of the boards are laid in place and checked for alignment, glue and nail the assembly together, and paint it. Then add a drain hole—between the buckets and the wheel's center—to release any water that manages to sneak through to the interior of the spinner.

Next, assemble the axle for the unit. Fasten a 3/4″ floor flange to each of the wheel's sides—at the center points—with four threaded rods and nuts, then slip a 3/4″ shaft of round steel stock through the wheel. Support the rod with two sealed 3/4″ pillow-block bearings secured to the structure's frame. Between the bearing and the flange on one side of the wheel, insert a threaded 3/4″ to 1/2″ brass male-to-female reducing bushing. (Bore out the inside threads of this piece so the shaft will run through the bushing's center.) On the other side of the wheel, weld a 3/4″ X 2″ pipe nipple, which screws into one of the floor flanges. This last addition allows the wheel to be removed easily from the shaft for routine maintenance and repairs.

The frame itself can be welded together out of scrap pieces of angle iron. Its shape isn't critical, but do make sure the support structure is stout enough to hold the wheel assembly above the streambed.

The shallow-well pump can be attached at a point near the base of the frame. Then, in order to gear the pump/wheel assembly to insure that each revolution of the waterwheel forces several back-and-forth strokes of the pump's single piston, install a 9″ pulley on the waterwheel's shaft, mount a little 2-1/2″ pulley on the double-valve pump, and connect the two with a V-belt of appropriate length.

Now erect a simple wooden box flume—with a hole cut out of its bottom to serve as a spillway—over the wheel, and hook up the 1″ inlet and outlet drinking-water lines to the pump. Then attach the 1-1/2″ "dirty water power" lines to the flume's entrance, and the simple machine—like the wheels of a slow-starting locomotive—will begin to turn.

The diminutive waterwheel requires little care. Once every few weeks, check the tension on the fan belt, clean any debris out of the screens at the top of the wheel's "power pipes", and lubricate the pump and bearings.

The project is both uncomplicated and inexpensive, if you can locate a small cast-aside or secondhand, shallow-well, double-action pump. These little water-pushers used to be quite common, but the few new ones currently available are costly, often carrying price tags in the hundreds of dollars. Abandoned pumps can frequently be obtained at a local junkyard, though, for about a tenth of the price of a new unit or possibly for less.

In short, this water-moving system is a reasonably priced, low-technology setup that could help anyone who has a source of usable "power" water, a supply of fresh drinking water, and the need to give the potable liquid a bit of an uphill boost. The small-scale waterwheel doesn't score high marks in mechanical efficiency, but its very simplicity makes it so easy to construct. And make no mistake about it, the water pumper does work. Its virtues are not all practical ones, however. One of the most appealing assets of the tiny waterwheel is its attractive, picturesque appearance.

BLOWING IN THE WIND

Windpower was used extensively to pump water and to produce electricity on farms and homesteads in the United States and Canada before the wire tentacles of the electric utilities reached into these rural areas. And wind-driven generators began to enjoy a resurgence in the 1970's when broad-scale rising energy costs pushed up the price of electricity.

Interest developed in restoring old direct current generating plants, and people began experimenting with alternating current powerplants and with alternatives to the traditional horizontal-shaft windspinners.

Of course, the early days of windpower in this country saw a lot of experimentation on the parts of the pioneers in the field, and this led to the development of efficient machines and to the solution of many of the problems exposed along the way. In the pages to come, Marcellus Jacobs, whose many windplants have been in operation for decades in all parts of the world, outlines some of his experience in the field.

But the wind is not limited to such major tasks as pumping water and producing electricity. It can be used on a much smaller scale to perform specialized tasks such as washing clothes.

BLOWING IN THE WIND

**The Wind Generator
Electricity From Thin Air**

The old-fashioned windmill with its many blades was a common sight on farms in the midwestern plains, but while the configuration of these windspinners was well suited to pumping water, it was not especially good for generating electricity. A three-bladed spinner seems ideal for a wind-driven powerplant since such a rig efficiently catches both light breezes and heavier winds.

For people in remote areas or for those who would like to cut the steel umbilical cord that ties them to the electric utility company, harnessing the power of the wind to light their homes, cook their food, and run their appliances has a certain obvious appeal. Wind-driven powerplants capable of serving a farm or homestead are nothing new to this continent, and the wind chargers were once quite popular, especially in the rural plains areas of the United States and Canada.

The first wind generator was probably built by the Arctic explorer Fridtjof Nansen in 1894. Ice-bound in the polar seas, Nansen rigged up a Holland-type windmill which drove a dynamo that charged batteries. He was enjoying electric lights in the vicinity of the North Pole when the houses of New York, London and other major cities in America and Europe were still illuminated with kerosene and gas lamps.

Marcellus Jacobs began production of wind-driven electric plants for use by farms and homesteads far from power lines in 1933, and the Wincharger Corporation of Sioux City, Iowa followed suit a few years later. Wind generators were

TURNING THE WIND INTO WATTS

used extensively in some rural areas up until the early fifties, when the Rural Electrification Administration brought power to most of the country. The Wincharger Corporation finally ceased production on all models except a small 200-watt unit in 1953, and the Jacobs Company stopped building its wind generator in 1956, but both are back in limited production today.

Because of the energy crisis and concern over air pollution, more and more people are looking for alternative power sources. Do wind generators offer a valid solution? The answer is both yes and no. To understand such a seemingly ambivalent reply, one first must understand some basic facts of electricty.

The three most important units of electrical measurement are volts, amps, and watts. The easiest way to remember the difference between amps and volts is to think of amps as current which is measured in much the same way as the volume of water in a pipe. And if amps are volume, then volts can be thought of as pressure, or the amount of push behind the volume of water. This is an important distinction: There's a world of difference between the flow of a swift mountain brook (high voltage, low amperage), and the flow of sluggish water in a large, slow-moving river (high amperage, low voltage).

Obviously, the relationship between amps and volts is very important, and the combination of the two when multiplied is the amount of electricity available. This total is measured in watts. So, when considering the installation of a wind generator, a question must first be asked: How many watts does it produce?

A 200-watt generator might serve a couple living simply in a small dome or one-room cabin and requiring only lights, but it would hardly be adequate for a homesteading family that would like to use electric lights plus a freezer, a stove, an arc welder, and power tools.

The easiest way to visualize the output of any given generator is to imagine the number of 100-watt bulbs it will fully light at one time. A 200-watt generator will handle two 100-watt bulbs, and a 3,000-watt unit will simultaneously power thirty 100-watt bulbs. To find out how much electricity is needed, just add up the total wattage of all of the light bulbs and appliances you plan to use, and tack on a few more for good measure.

Once a person's power needs are known, it's time to determine what kinds of wind generators are available and where to get one that will meet the energy demand. There are three ways to go: a new wind generator, a used powerplant, or a homebuilt unit constructed from automobile components.

Unless you can afford to spend several hundred dollars, perhaps more than a thousand, low-wattage new wind generators are out of the question. And, until someone goes into production on a high-wattage machine at a reasonable price, most people must buy a used unit or build their own.

Wind generators were common sights on farms in the Great Plains states during the 1930's and '40's, and while a few may still be standing, most of the machines were taken down and sold for scrap after the REA reached these rural areas with its electric lines.

The two most common wind generators were the Wincharger and the Jacobs. The Wincharger came in several models, from 6 volts to 110 volts and from 200 watts to 1,200 watts. The Jacobs, a much heavier machine, was built in 32-volt and

BLOWING IN THE WIND

The Wind Generator
Electricity From Thin Air (continued)

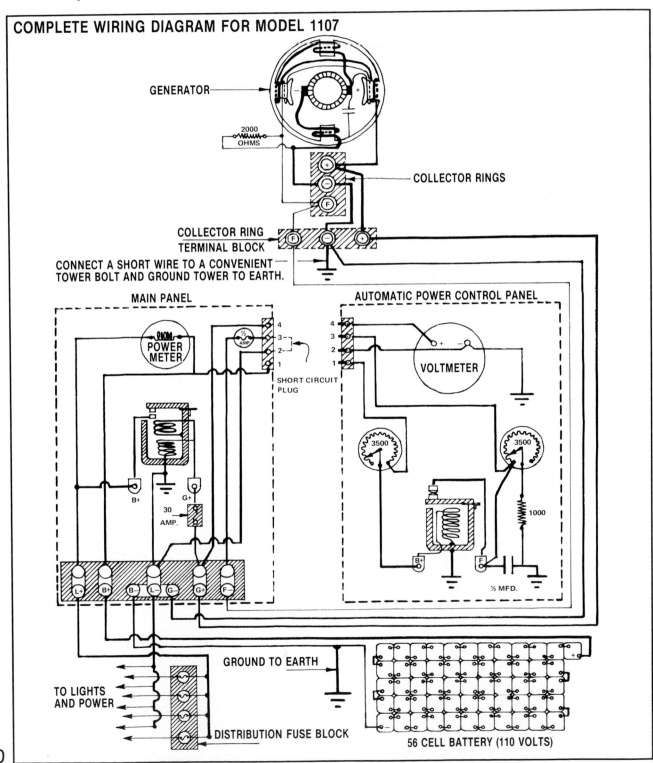

COMPLETE WIRING DIAGRAM FOR MODEL 1107

GENERATOR

2000 OHMS

COLLECTOR RINGS

COLLECTOR RING TERMINAL BLOCK

CONNECT A SHORT WIRE TO A CONVENIENT TOWER BOLT AND GROUND TOWER TO EARTH.

MAIN PANEL

AUTOMATIC POWER CONTROL PANEL

POWER METER

½ AMP

SHORT CIRCUIT PLUG

VOLTMETER

3500

3500

B+

G+

30 AMP.

1000

L+ B+ B− L− G− G+ F−

B+

F−

½ MFD.

GROUND TO EARTH

TO LIGHTS AND POWER

DISTRIBUTION FUSE BLOCK

56 CELL BATTERY (110 VOLTS)

TURNING THE WIND INTO WATTS

110-volt models that ranged from 1,500 to 3,000 watts. In their day, the 32-volt units of both makes were the most popular. Of the two, Wincharger is the brand you're more likely to encounter now. The Jacobs is quite scarce.

Almost any old wind generator you may locate will probably need extensive restoration. It's unusual to find one that still has usable propellers, which being made of wood (with the exception of some later models of the Wincharger, which had aluminum props) are the first parts to deteriorate. After all, the machine has probably stood untended for well over 25 years of summer thunderstorms and winter blizzards. Finding a generator that still has the original control box is even rarer, though one can be fabricated by almost any electrician.

Whether restoring one of these old wind-electric plants is worthwhile depends on finding a machine that doesn't have extensive damage, such as broken castings or missing major parts.

Once the generator has been purchased, the first job is to get the thing down. Unless you're a telephone lineman or a circus tightrope walker, working on top of a tower is a scary proposition. However, once you've tied yourself on with a safety belt, it isn't so bad, except when the wind starts to blow and the generator wants to move around with it. (Even with the vane turned "out of the wind" so that the tail is perpendicular to the propeller shaft, the mechanism will want to turn when the breezes get gusty.) Do not delude yourself into thinking that the tower can be lowered safely with the generator still in place.

For Winchargers, a hoisting device (constructed as indicated in the upper diagram on page 112) is securely bolted to the tower with enough room above the generator to lift it free of its mount with a block and tackle. Propellers, vane, and any other easily removable parts are taken off and lowered with the help of an assistant on the ground who handles the block and tackle. The generator itself will, of course, be the last and heaviest load.

However, this type of hoisting pipe is totally inadequate for removal or installation of a Jacobs machine, which is much heavier and will invariably bend the homemade implement. The Jacobs company has suggested that you enlist the help of a nine-foot length of well-drilling pipe about two inches in diameter. No crossarm is used, and the hook for the block and tackle is simply placed in the upper opening of the tube.

Bear in mind that many of the nuts and bolts you'll be working with are likely to be quite rusty. A can of Liquid Wrench or a similar product will help in loosening them. Parts under tension require the most care, and it's a hair-raising experience to have a spring-loaded tail vane pop loose on you when there's no place to duck. Anyone working around the tower should wear a hard hat.

For carrying tools, a carpenter's or a lineman's tool pouch would be a good investment. It's most inconvenient to have your pockets bristling with screwdrivers, wrenches and hammers, and it's bothersome to have to worry about the possibility that a tool may fall from your pocket and injure someone on the ground.

Once the generator is off the tower, it's ready to be hauled home and torn down to see what makes it tick. In addition to the generator, be sure to take the "stub", which is the top five feet or so of the tower. This contains the generator's pivoting mechanism and the all-important slip-ring collector which allows the generator to turn in any

BLOWING IN THE WIND

The Wind Generator
Electricity From Thin Air (continued)

direction without twisting the wires which lead to the battery bank.

Also, be certain to ask the former owner for any wires and control boxes that go with the generator. With luck, he'll have them stashed away in the toolshed, but if not, an electrician can be hired to make a new control box. If your generator is a 32-volt model, and you don't plan to install an inverter to convert 32 volts DC to 110 volts AC, remember to ask the seller if he has any old 32-volt DC motors lying around. He'll have no more use for them, but you'll certainly find them handy.

Rebuilding, or in most cases *making*, a wind generator prop is not as difficult as it may seem at first. A crude one can be hacked out of an old 2 X 4, and it will turn up a storm when faced into the wind.

The control box poses another problem, but with the help of an old wiring diagram from the 32-volt Wincharger manual, a qualified electrician should be able to put one together without much trouble. A "control box" of sorts can be as simple as the one shown in the lower diagram on this page, which really amounts to an antimotor device and protection from lightning. (A DC *generator* will act as an electric *motor* if the current is reversed. Without an antimotor device, the juice from the batteries will "motor" the generator when the wind isn't blowing, and this will eventually discharge the storage units completely.)

A lot of different opinions exist on the subject of what kind of batteries are best, but the kind of storage cells used by the old-timers were large industrial-type 2-volt units connected in series to add up to the voltage of the generator. In most cases, that was 32 volts, or sixteen 2-volt cells hooked up to make one big 32-volt battery.

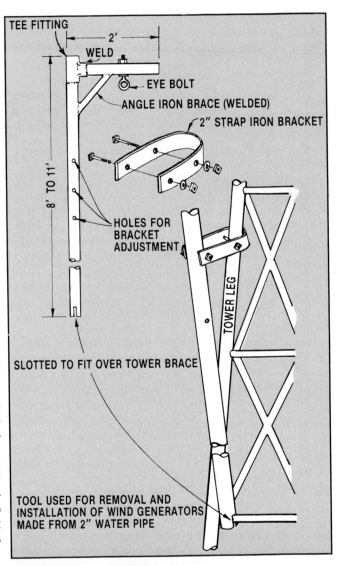

TEE FITTING
2'
WELD
EYE BOLT
ANGLE IRON BRACE (WELDED)
2" STRAP IRON BRACKET
8' TO 11'
HOLES FOR BRACKET ADJUSTMENT
TOWER LEG
SLOTTED TO FIT OVER TOWER BRACE
TOOL USED FOR REMOVAL AND INSTALLATION OF WIND GENERATORS MADE FROM 2" WATER PIPE

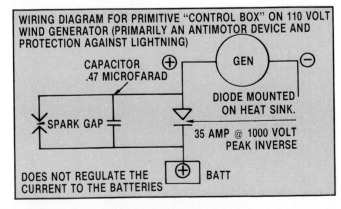

WIRING DIAGRAM FOR PRIMITIVE "CONTROL BOX" ON 110 VOLT WIND GENERATOR (PRIMARILY AN ANTIMOTOR DEVICE AND PROTECTION AGAINST LIGHTNING)
CAPACITOR .47 MICROFARAD
GEN
DIODE MOUNTED ON HEAT SINK.
SPARK GAP
35 AMP @ 1000 VOLT PEAK INVERSE
DOES NOT REGULATE THE CURRENT TO THE BATTERIES
BATT

TURNING THE WIND INTO WATTS

These units are still available from the industrial division of any battery manufacturer. They're heavy and expensive, but in the long run are probably the most efficient and economical way to go. Some units have a life expectancy of 25 to 30 years, and prorating their cost over that period of time shows the expense to be quite reasonable when compared to what you'd be paying for electricity. The minimum capacity a person should consider for efficient operation is 180-amp-hour batteries. (The 110-volt Wincharger manual speaks of batteries in the 240- to 424-amp-hour range.) If the price of the big units is prohibitive, golf cart or diesel truck batteries will be a better bet than ordinary automotive units.

People who can't afford a new wind generator, can't locate a used one in restorable condition, and really don't want electricity for much more than a small stereo set and a few 25-watt light bulbs might still want to consider fabricating a homemade windplant. If you're a do-it-yourself type who enjoys tinkering with mechanical devices, you can build your own wind generator out of automobile parts, though the highest output you should expect from such a design would be about 600 watts.

In recent years, several different designs for homebuilt wind-driven generators have appeared, and while some of these schemes aren't worth the time and money required to build them, many have promise for folks with some basic tools and the skill to use them. One do-it-yourself system is based on the very rare Jacobs Model 15, which was a 1,500-watt flywheel-driven unit that sported two 750-watt generators. Surplus high-wattage aircraft generators are utilized to bring potential power output to 4,800 watts or more! With a well-designed propeller, a homebuilt plant could come close to delivering 300 kwh a month in an area with an average wind speed of 10 MPH.

Another source of hope and encouragement for folks yearning to be energy self-sufficient is the recent proliferation of new ideas in the field. In fact, the days of the horizontal-axis windplant may be threatened by the introduction of vertical-axis machines. The design has many advantages over the more traditional "windmill" type, most notably, the easier and less expensive construction of the blades (usually the most complicated part of a do-it-yourself unit), the lack of a need for a tail vane (because the generator accepts wind from any direction), and the fact that the machine can be mounted at ground level (which means a lot to anyone who's not the tower-climbing type!).

The drawbacks to a vertical-axis generator like this are the dangers inherent in the whirling blades being nearly at ground level and the loss of the higher velocity winds found at 30 or 40 feet up.

Whether a person finds a used Jacobs or Wincharger, buys a new unit, constructs one from surplus aircraft or automotive components, or waits around for someone to manufacture a newly designed high-wattage generator for a reasonable price, there is one very important consideration: Before devoting a lot of time, energy, and money to any wind-electric system, be sure you have gathered accurate data on the average winds in your area.

If your location's wind speed averages less than 10 MPH, you'll need to think of other ways to get your electricity. Don't let your enthusiasm for an alternative power source blind you to the realities of nature's laws.

BLOWING IN THE WIND

A Wind Monitor

The first and most important step in harnessing the wind is deciding where to place your equipment. Selecting the right spot for a windplant—or even just choosing a suitably sized generator—can be nearly impossible unless you know how much wind to expect and from what direction it's likely to come. After all, the amount of wattage that a wind-driven powerplant will produce is actually related to the cube of the wind velocity, and thus a small difference in speed can make a big change in the amount of electricity that can be generated. (For example, a 15-MPH breeze will actually yield about twice the energy of a 10-MPH puff.)

Of course, there are a number of commercial wind-monitoring systems available, and—while many of them are excellent products—they tend to be quite expensive. So why not build a wind gauge and vane for yourself?

The two parts of this system have many similarities in their construction, and both can be mounted on the same PVC pipe stand. The anemometer, however, is just a bit more complicated than the wind vane.

The heart of this homemade anemometer is a small electric motor—with permanent magnets and windings—that can also operate as a generator. It's often possible to find excellent examples of anemometer-sized motor/generators in children's discarded toys if you don't wish to purchase a unit.

To turn the little motor into a generator that'll give a current proportional to wind speed, you'll need to give the powerplant "wings" to catch the breeze. A DC ammeter will then measure the motor's output, and you can calibrate the gauge to read in MPH.

The powerplant needs to fit snugly within a 3/4" to 1/2" Schedule 40 PVC cup reducer, but you'll probably have to cut notches in the fitting's seat to accommodate the wire tabs that emerge from the motor/generator's bottom. A 3/4" Schedule 40 PVC coupling can then be slipped over the motor and cup reducer, enclosing the rest of the generator. Now, grind the tips of the eight-sided plastic fitting down until they're flush with the 3/4" coupler, cement the unit into the PVC housing with all-purpose glue, and seal the assembly with silicone adhesive, being careful to avoid getting any of the sealant on the axle.

To keep water from leaking in around the motor/generator's shaft, cut a 1/4"-diameter washer from a piece of felt, and poke a hole in the center of the material with a needle. Slip the washer over the generator's shaft, then oil it with sewing machine lubricant.

Two bisected table-tennis balls serve as the wind-catchers that spin the homemade gauge. The spheres are most easily divided by cutting around each one's seam with a razor knife. Don't puncture the surface: Just make a progressively deeper cut until you can split the ball by squeezing it gently.

The hemispherical sails are connected to the central spinner by 3-1/2"-long sections of coat hanger. To mount the rod into each cup, snip two notches in the lip of the plastic—directly opposite each other—and glue the piece of coat hanger in place.

While the adhesive is setting, locate a 1-1/2"-diameter, 2-1/2"-long prescription bottle. Drill four 7/64" holes—spaced at 90° intervals around the circumference—into the sides of the vial, at

TURNING THE WIND INTO WATTS

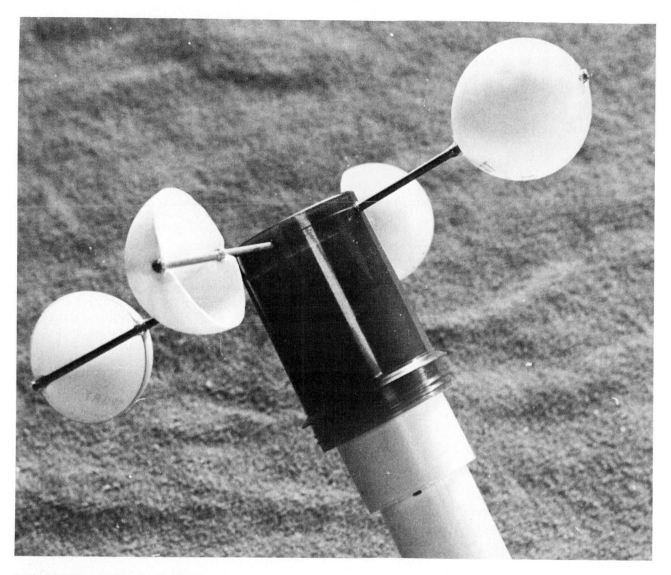

Cup-shaped propellers made from
halves of table-tennis balls catch
the wind and spin a small direct-
current motor. This motor acts as
a generator and produces a current
proportional to wind velocity.

BLOWING IN THE WIND

A Wind Monitor (continued)

points just above the bottom. Then bore a 5/64″ hole exactly in the center of the bottle's base, using the bottom's casting "tit" as a guide for positioning the drill bit.

Next, turn the vial upright and insert a "sail arm" into each of the 7/64″ holes. Position the table-tennis-ball halves so that they'll catch the wind (either clockwise or counterclockwise motion will do). Set them so that each one is an equal distance from the bottle, but so that the rods don't obstruct the central 5/64″ hole.

Epoxy will hold the coat hangers in place within the container, but before you pour the adhesive into the vial, you should place your 5/64″ drill bit (coated lightly with oil so the epoxy won't stick to it) into the 5/64″ hole to keep a central opening for the motor/generator's shaft. After drilling a corresponding hole in the bottle's top—to keep the bit lined up—mix up a small amount of fast-setting epoxy, pour it in 1/4″ deep, and then slip the lid into position to support the "far end" of the drill. After five minutes of setting time, gently remove the drill and put the assembly aside to finish drying completely.

On the morning after pouring the epoxy, you can attach the spinner to the motor. Just slip the prescription bottle unit over the generator's shaft—the fit should be snug—and test the device for balance and squareness by spinning it. Slide it all the way down on the motor's shaft, apply a drop of cyanoacrylate glue to the tip of the shaft, and then slip the vial back out until it's flush with the end of the shaft.

To wire up the anemometer, solder a 1,000 (1K) ohm calibration potentiometer across the 0-1 milliampere meter's terminals, using only the central wiper terminal (the lone one, opposite the other two) and either of the remaining connectors. Now wrap each wire from the "sender" around a terminal on the meter (don't solder yet). Then turn the potentiometer—using a small screwdriver—all the way in one direction and give the sender a spin. If the needle fails to move, turn the control all the way in the other direction and repeat the test. A healthy spin should now produce about a half-scale reading. If it does, you can complete the construction by soldering the wires to the terminals on the meter.

A well-made commercial anemometer will provide you with an extremely accurate basis for calibrating your device, but you can make adequate adjustments using your car's speedometer. Temporarily mount the sender atop a five-foot piece of plastic pipe—no glue is necessary at this point—and have a friend pilot the family carriage while you hold the spinner in the wind. Have your driver proceed down a level, deserted section of road at 25 MPH, so that you can adjust the control to produce a meter reading of "1". Make a couple of passes back and forth to compensate for the wind direction and for your car's streamlining and speed inaccuracy. Then have your assistant drive the car at 20, 15, 10, and 5 MPH so you can note those readings. (You can later make a new faceplate for the meter to indicate the correct wind speed in MPH.) And once you've mounted the meter in the center of a 12″-diameter wooden disk, your anemometer will be complete.

The weather vane portion of the homemade wind-monitoring center consists of a sail, which is connected to a rotary switch that, in turn, is wired to a series of lamps arranged in a circle on the board. As the wind direction changes, the

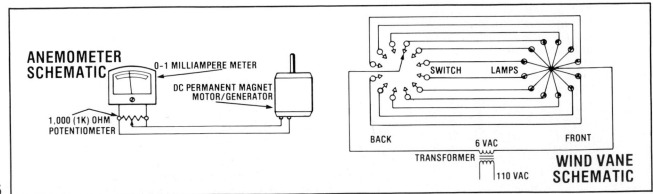

ANEMOMETER SCHEMATIC — 0-1 MILLIAMPERE METER — DC PERMANENT MAGNET MOTOR/GENERATOR — 1,000 (1K) OHM POTENTIOMETER

SWITCH — LAMPS — BACK — FRONT — 6 VAC — TRANSFORMER — 110 VAC — WIND VANE SCHEMATIC shaft will rotate and touch different contacts—thereby lighting corresponding lamps—which then indicate each new direction of the wind.

The mechanical components of this instrument are quite similar to those used in the building of the anemometer: a coat hanger, a plastic vial, and PVC pipe parts. The wind-catching portion of the instrument, however, is formed from steel rather than from table-tennis balls. The best free source of suitable metal is a discarded steel beverage can.

Using an opener, remove the top and bottom of the can and then cut along its seam with a pair of tinsnips. Before you attempt to flatten the metal, snip away the "lips" where the lids were joined to the can's body.

At this point you'll notice that the steel still doesn't want to stay flat. First, use the tinsnips to cut the metal to the shape shown in the photograph on page 118. Then, in order to strengthen the flimsy sheet, form three lengthwise ridges in the piece by laying the vane on a length of coat hanger and sliding a short section of 2 X 2 along the metal directly above the wire. The pressure will create grooves in the tailpiece.

Now, cut a straight 12″-long section of coat hanger and remove its paint with sandpaper. Once that's done, fasten the support rod into the center groove in the vane, using acid core solder and a torch.

In order to avoid having the tail's weight put lateral pressure on the bearing, you'll need to build a counterbalance. You can fashion one by filling the decorative endpiece of a curtain rod with solder and plunging a short piece of coat hanger into the still-molten lead. (The wire must, of course, be sanded and tinned if it's to bond to the solder correctly.) A number of other approaches could also be used to provide the needed counterweighting, such as welding nuts or washers to the coat hanger section.

Whatever approach you use, the tail and counterweight must be mounted into the housing—which is formed from another 1-1/2″ X 2-1/2″ prescription vial—to establish perfect balance. Drill two 7/64″ holes (one on each side of the bottle) to accept the "free" ends of the coat hanger pieces and balance the rods in the openings, adjusting the length of the counterweight "stem" to level the assembly.

The switch shaft will be set into the pill bottle in much the same fashion that the motor shaft was mounted for the anemometer. Drill a 15/64″ hole in the casting tit on the bottle's bottom, place the oil-coated drill bit into position, and pour the epoxy in around it to a depth sufficient to secure the tail and counterweight rods.

A 14- or 16-wire telephone cable is the easiest material to use for wiring the rotary switch. Solder one wire to each of the 13 terminals on the bottom of the switch. Be certain to note the colors of the wires as they are soldered sequentially around the 12 outer terminals of the switch.

Now mount the switch in a 3/4″ Schedule 40 PVC pipe coupler. Unfortunately, the oval shape of the switchplate is slightly larger than the coupler's inside diameter, so you'll have to cut (with a hacksaw) a 3/8″-wide, 5/8″-deep notch on each side of the coupler to provide clearance. Furthermore, to reduce the friction of rotation, you should remove the index ball located under the switch's metal mounting bracket.

Once you've finished the wiring and mounting and have allowed the epoxy to set overnight, you

A milliampere meter is wired to indicate wind speed while the direction that its blowing is shown by the lights around the gauge. The generator for the anemometer is mounted in a PVC pipe fitting.

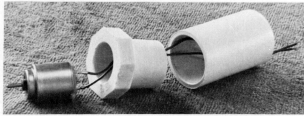

BLOWING IN THE WIND

A Wind Monitor (continued)

can begin putting the pieces together by threading the cable into the coupling and slipping the switch into its grooves. When the device is seated, check to see that the shaft is aligned accurately, or wobbling and sticking will result.

To seal the housing, wrap electrical tape around the upper (cutaway) portion of the coupler, and fill the recesses with silicone sealant to protect the contacts from moisture. (Don't *force* the adhesive into the switch holes, or you'll run the risk of hindering the unit's operation.) While you have the caulk in hand, seal the other end of the switch housing, but avoid getting any material on the shaft of bearing.

Then, to complete the vane assembly, slide the vial/tail/counterbalance body over the switch shaft and down until it contacts the parts inside. Then place a drop of cyanoacrylate glue on the end of the shaft, slip the vial back out until it's flush with the shaft end, and allow the adhesive to dry.

Twelve lamps are arranged in a circle—like the hours on a clock—to serve as the wind direction readout. Although you could use a number of different kinds of lamps, it's easiest to use readily available units that come with attached leads to facilitate wiring.

In the face of the board on which the wind speed indicator is mounted, drill a dozen holes large enough to accomodate the bulbs. These holes should form a ring around the meter. Next, slip the lamps into the openings, and connect one lead from each bulb to the power source. Then solder the other wire from each light to one of the switch's contacts. Refer to the color code chart and connect the lamps in sequence to the wires from the switch contacts. It doesn't matter

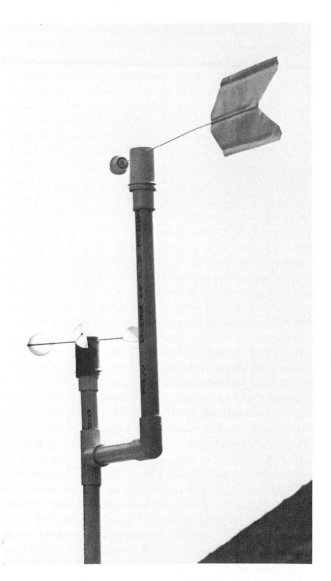

Both the anemometer and the wind vane can be mounted on the same length of pipe, but the directional indicator should be offset and well above the cups.

118

TURNING THE WIND INTO WATTS

where you start, but you must maintain the proper order. Finally, attach the wiper lead from the switch to one leg of your power supply.

The wind vane can be mounted by teeing a second vertical piece of pipe from the anemometer mount, but be sure to set it a foot or so above the anemometer to avoid having one device interfere with wind flow to the other.

Before you glue the wind vane atop the 3/4″ PVC stand, you must orient the assembly. First turn the vane so it's pointing due north. Then just have someone check the display to see which lamp is lit. Simply rotate the switch housing (without turning the vane itself) until the north lamp lights up, and then glue the coupler to the pipe in that position.

Move your wind vane and anemometer to different locations periodically, and record—on a regular basis—the readings you get at each site. You should note the speed and direction of the wind at least twice a day over a two-week period.

Start by checking hilltops, and then move on to fields that are unobstructed by trees, hills, or buildings. (If the equipment is set farther than 100 feet from the meter and lamps, use at least 18-gauge wire to prevent line losses.) In general, a good windplant site will have breezes of more than 10 MPH three times a week or more. But keep in mind that one 20-MPH blow will produce more power than three 10-MPH zephyrs. And keep in mind that wind velocity is 20% to 50% greater 30 or 40 feet above ground level.

If you keep careful records, your wind-energy-measuring equipment will help you to make the best possible decisions about where you should put a windplant and how large a generator you will actually need.

LIST OF MATERIALS

ANEMOMETER

- (1) Small DC electric motor
- (1) 0-1 milliampere meter
- (1) 1,000 (1K) ohm control
- (1) 3/4″ to 1/2″ Schedule 40 PVC cup reducer
- (1) 3/4″ Schedule 40 PVC coupling
- (1) Coat hanger
- (2) Table-tennis balls
- (1) 1-1/2″-diameter, 2-1/2″-deep prescription bottle
  - Fast-setting epoxy
  - 3/4″ Schedule 40 PVC or steel pipe
  - Hookup wire
  - Silicone seal
  - All-purpose glue
  - Felt washer
  - Cyanoacrylate glue
  - Sewing machine oil

WIND VANE

- (1) 12″-diameter disk made from lumber or plywood
- (12) 6-volt lamps
- (1) 6-volt transformer
- (1) 12-position rotary switch
- (1) 1-1/2″-diameter, 2-1/2″-deep prescription bottle
- (1) 3/4″ Schedule 40 PVC coupling
- (1) Coat hanger
- (1) Steel beverage can
- (1) 3/4″ Schedule 40 PBC "T"
  - 14- or 16-strand hookup wire
  - Fast-setting epoxy
  - 50/50 solder
  - Silicone seal
  - Electrical tape
  - Cyanoacrylate glue

BLOWING IN THE WIND

A New Twist
The Savonius Rotor

With a revival of interest in energy sources other than utility companies, many Americans are reconsidering the generation of electricity from the wind's power. Until recently, however, only one type of wind charger was readily available: the propeller-driven generator or alternator.

The props that spin the generators on conventional windplants vary in number of blades (two, three, or four) and in the complexity of their aerodynamic surfaces. The differences notwithstanding, the price for even an owner-built prop-driven mill would strain many a homesteader's budget.

But experimentation has led to a little-known alternative wind device which is low in cost, is simple to construct, and boasts several other distinct advantages over a conventional mill, both in general performance and in safety of operation. This unit is the Savonius rotor: It's often called the S-rotor because of its appearance.

A Savonius rotor is easily made: Just split a cylinder equally through its length, offset the halves by a distance equal to the radius of the original form, and secure the segments to end plates that are the width of the new diameter. Then insert a rod through the center of the assembly and fix its ends in bearings, and the device will rotate when exposed to the wind. If you use soft drink, beer, or other small cans for your rotor, you'll have a toy. But, if you start with 55-gallon drums stacked three on end (out of phase with one another), even at low wind speeds the power from your creation will surprise you.

You'll be even more surprised by the S-rotor's performance when it's compared to that of the prop-powered mill. If both are tested in wind tunnels, the Savonius design appears inferior, but under normal outdoor conditions the results are almost reversed. To see why, some understanding of the nature of moving air masses is needed.

There are two basic types of wind: prevailing and intermittent. The former blows an average of five days out of seven in a given area, and the latter only two. While intermittent winds occur only 35% of the time, they provide 75% of the power available from moving air masses.

Intermittent winds come mainly in the form of gusts which "ride on" a prevailing breeze but usually deviate from it in direction by 15 to 70 degrees. The practical importance of this fact can be demonstrated by placing a conventional mill and an S-rotor side by side in a steady wind. Suddenly there's a gust, and the propeller-driven unit swings into it. Then, as the puff dies, the windplant's tail slowly moves the fan back into the prevailing wind. The S-rotor, meanwhile, just speeds up in the rush of air and slows down as the velocity drops.

Here's the point: The prop-driven mill needed time to align itself first with the gust and then with the steady wind and as a result could not take advantage of all of their force. The S-rotor—which didn't have to swing or "track"—was able to absorb the full power of both. One of the Savonius design's greatest assets, in fact, is that it can take a wind from any direction at any time.

That same characteristic also gives the S-rotor a great advantage in durability. In a steady, low-speed wind the swinging of the prop-powered mill is mechanically acceptable, but at higher speeds it's another story. The spinning propeller is just one big gyroscope, and its constant adjustment to the direction of the moving air exerts tremendous forces. The resulting "gyroscopic vi-

TURNING THE WIND INTO WATTS

bration'' has sent many a propeller, generator, and tower crashing to the earth. The nontracking Savonius unit experiences no such problems.

Even if an S-rotor did break loose, it wouldn't have far to fall. Unlike the conventional windplant—which rotates horizontally and is mounted, along with its generator, on top of a tower—the Savonius device spins about a vertical shaft. This means that its alternator can be mounted on or near the ground. The tower is replaced by a pole with some guy wires. Think about that: easy access to the alternator, easy lowering of the unit, and easy relocation, all without an elaborate supporting structure!

With the two units in action side by side, another major difference is apparent: The S-rotor appears to turn very slowly, with only one revolution for up to eight of the conventional mill's. However, if you think that speed is necessary to a wind charger's function, think again. True, the usual windplant must attain a high RPM to operate, but the S-rotor—which presents 10 to 20 times as much surface area to the moving air mass—develops the same amount of power at lower rotational speeds.

Also, since the Savonius plant's rotational speed is relatively low, its power output must be stepped up through some rather high gear ratios to drive an alternator fast enough to produce a meaningful amount of electricity. But so what? Such gearing creates no restarting problems for the S-rotor (as it does for a propeller-driven unit) and is entirely practical for the Savonius machine.

The S-rotor has yet another advantage over the propellered mill. The faster-turning blades of the latter must be designed and balanced precisely

The Savonius rotor is low in cost, is simple to construct, and boasts distinct advantages in general performance and safety of operation.

BLOWING IN THE WIND

A New Twist
The Savonius Rotor (continued)

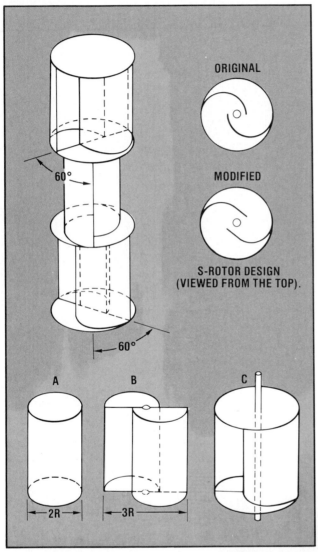

ORIGINAL

MODIFIED

S-ROTOR DESIGN
(VIEWED FROM THE TOP).

A   B   C

2R   3R

to operate at high speeds. Since few folks have the tools or know-how to do this, the airfoils (or the whole propeller) must often be purchased at high cost. By contrast, the slower-moving S-rotor needs minimal or no balancing, and its "wings" can be constructed quite simply and easily.

Then there is the question of "feathering", or twisting, of a conventional windplant's propeller blades to lower the device's rotational speed in high winds. Feathering is a necessary precaution with most wind chargers for a number of reasons. First, whenever a fixed gear ratio is used, some such governing arrangement is needed to prevent the alternator/generator from exceeding its maximum rated output (current). In a propeller-driven unit, however, the prop itself is a factor in determining at what windspeed feathering must occur. The problems involved may be described in terms of balance, structural design and blade tip stresses.

Balance is always more important (or critical) at higher rotational speeds than at lower RPM. If a propeller is allowed to exceed its "operating range", dangerous vibrations can be set up. The structural design of a propeller and the materials of which it's made also determine the device's upper rotational speed beyond which centrifugal force will pull it apart. Tip stresses occur at higher wind speeds, but are rather independent of whether or not the propeller is turning. This is one reason why "braking" a fan to stop in a tempest won't help it very much.

All these problems of propeller-driven units are irrelevant to the S-rotor. Since the Savonius device never achieves high rotational speed, balancing is not critical and centrifugal force won't pull the device apart. In addition, a runaway S-ro-

(A) original cylinder (with a diameter of 2R); (B) offset, split cylinder halves; (C) end caps of 3R diameter fitted to drum halves. Three of these stacked (as tiers) and 60° out of phase comprise a completed S-rotor wind assembly.

S-rotor Electrical Plant—One design, allowing the S-rotor (and its gear) to turn about a fixed shaft. This requires the alternator to be positioned immediately below the bottom end cap. Spoiler and wind-sensor (for activation of field current to the alternator) are not shown in this drawing.

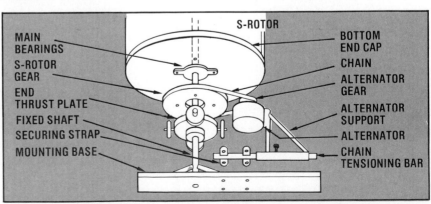

MAIN BEARINGS
S-ROTOR GEAR
END THRUST PLATE
FIXED SHAFT
SECURING STRAP
MOUNTING BASE

S-ROTOR

BOTTOM END CAP
CHAIN
ALTERNATOR GEAR
ALTERNATOR SUPPORT
ALTERNATOR
CHAIN TENSIONING BAR

TURNING THE WIND INTO WATTS

tor will soon begin to spoil (or brake) itself because its own wings get in one another's way at higher RPM.

Finally, there's one more feature of the Savonius unit which isn't apparent to the eye, but which has shown up in tests. The S-rotor can begin to charge 12-volt batteries at wind speeds lower than the 7-MPH minimum required by the "normal" prop-driven plant. Such units have been known to work successfully at 6 MPH, and some modification of the rotor may lower the necessary wind speed to 5 MPH. That could be a real benefit to people who live in areas of lower average wind speed, and who may not be able to use propeller-type systems.

The S-rotor design may be adapted to the opposite extreme, too, by installing a centrifugally operated spoiler to slow the rotation of an S-rotor in very high wind speeds. A separate sensor (also mechanical) is used to limit or switch off field current to the alternator in case of either very high winds or no wind at all.

ALTERNATOR(S)	MAXIMUM NORMAL OUTPUT (in watts)	PEAK OUTPUT (in watts)
45A	500	540
60A	680	720
130A	1,400	1,560

The previous figures show what kind of power the Savonius rotor can produce. Such a windplant's capacity, of course, depends a good deal on your choice of alternator, so this chart shows results for units with three ratings: a 45-amp (very common), a 60-amp (a little harder to locate), and a 130-amp (found in commercial vehicles or in luxury passenger autos). Included are figures for the peak output the plants can provide and for the normal maximum power.

Since an S-rotor can be built from readily available materials, the cost of one of these vertical axis units can be surprisingly low. In addition to the rotor itself, however, batteries and possibly an inverter are needed to complete a windpower plant designed around the Savonius windspinner. The latter, if required, is the same kind used in most conventional wind systems. This still leaves open the question of batteries for storing your power, however. Nickel-cadmium cells are close to ideal, though they are costly. However, if you can work a good deal with an auto salvage yard, you may be able to acquire enough lead-acid car batteries to do the job. At least ten 12-volt units are usually needed to store household power, and they should be tested for strength and durability.

Proper care and use of a lead-acid battery—if it's in good condition to start with—will give the storage cell many times the life of the same unit in an automobile. Even the costly nickel-cadmium batteries won't last long unless they're correctly charged, discharged, and maintained.

BLOWING IN THE WIND

Recollections of a
Windpower Pioneer

Marcellus Jacobs, one of the early pioneers in wind power, began development of his famous direct current generating systems in the 1920's. His powerplants were manufactured from the 1930's to the mid-1950's and proved durable beyond expectation. A plant installed in Antarctica was still spinning its blades and had most of its paint more than 20 years later, and other windplants of his have withstood hurricanes in the West Indies.

TURNING THE WIND INTO WATTS

One of the world's true pioneers, Marcellus Jacobs invented—almost single-handedly—the first practical wind-powered generating system and originated nearly all the noteworthy advances in the field from 1930 to 1956. By manufacturing and selling about 50 million dollars' worth of the units over a span of 25 years, Jacobs proved that windplants—*his*, at least—were a technically and economically viable source of electrical energy.

The windpower pioneer stopped manufacturing his windplants in 1956, but people who know his expertise are still fighting to find one of his old second- or thirdhand units. Why? Well, Admiral Byrd set up one of the systems at the South Pole in 1933, and on June 17, 1955 Richard E. Byrd, Jr. wrote a letter to Jacobs in which he said: "I thought it might interest you to know that the wind generator installed [by my father] . . . at the original Little America was still intact this year after almost a quarter of a century. . . . The blades were still turning in the breeze [and] show little signs of weathering. Much of the paint is intact."

Marcellus Jacobs, in short, designed good direct current windplants. He built them well, too . . . and he built them to last. Later, when there was a resurgence of interest in home-generated electricity, the pioneer returned to the field and designed alternating-current wind generators capable of being tied into the local power grids.

The following reminiscence is drawn from a 1973 interview with Jacobs. It offers some fascinating insights into both the man and the early history of windpower generation.

"After I left high school, about in 1920, I took one year of electrical training in Indiana and a special six-month course in electricity in Kansas City. Most of my education, though, just came from studying on my own. I got the books and picked up what I could from them, and thought the rest out for myself. When I was still in high school, I built and sold little peanut radios that operated on storage batteries . . . and pretty soon we wanted motors and welders and drill presses and what have you that operated on current. At the same time, I had always been intrigued by the wind. It was natural, I suppose, to put the two interests together.

"Our ranch was 40 miles from town, and in those days, of course, there weren't any Rural Electrification Administration lines running all over the country. We—there were eight children in our family—had to make do with kerosene lamps and so on . . . but we soon got tired of that. So we rigged up an old secondhand engine to run a little DC generator. However, it fluctuated every time the load changed, so we hooked up the generator to some old car batteries to balance the system and that worked pretty well. Along about then, though, we started a hand forge and put a motor on that, and we needed more current than our engine-driven generator would produce.

"I first tried to use a fan off one of the regular water-pumping windmills we had on the ranch. I took a Ford Model T rear axle and cut the side shaft off where one of the wheels was supposed to go, and I put the big fan on instead. Then I mounted the tail vane out where the other wheel should be, and I extended the drive shaft down to the ground where I had my generator. I just locked the differential with a pin so that as the wind turned the fan it would drive the shaft.

"But there were several things wrong with the setup. It wasn't efficient . . . you know . . . there was no real gain. One of those big water-pumping

BLOWING IN THE WIND

Recollections of a
Windpower Pioneer (continued)

windmill wheels is designed to catch all the wind in its diameter right at the start. Otherwise it'll just sit there. Unless the pump has lost its prime, that wheel has to lift water right from the instant it begins turning. It needs a lot of starting torque, and that's why it has so many large blades.

"Once the wheel gets up some speed, however, about 80% of those blades get in each other's way. They begin fighting each other. In fact, a water-pumping windmill needs all the power it generates just to run *itself* in an 18- or 20-MPH wind. You can pull the pump rod loose and the wheel won't over-rev and tear itself apart.

"The wheel we finally came up with for a windplant is altogether different. There's no load on it at the beginning, you see . . . just the very slight drag of two ball bearings. The three little blades sticking out of the wheel's hub are all you need to start the thing turning in a two-mile breeze. And those narrow blades are also all you need to catch every bit of air that moves through the wheel's diameter when the wind blows 20 MPH. They'll do it better than all those sails on a water-pumping windmill's fan, too. A three-bladed windplant propeller may develop between six and eight horsepower in an 18-MPH wind, while an ordinary windmill wheel of the same diameter sitting right beside it won't produce much over two.

"Well, we messed around for three years or so. We even made a governor that turned every one of the blades—to feather them—on such a wheel, but there were just too many other factors working against the design. To put it very simply: If you catch all the wind that moves through a certain diameter with three blades, there's no need to have fifty of them hanging out there. The extras just get in the way.

"I learned to fly in 1926 or '27, and that gave me the idea that an airplane-type propeller was what we wanted. Most of those, of course, had only two blades, so that's what we used. Actual airplane props didn't have the right pitch, but we made some windplant propellers that were quite similar to the ones used on aircraft. We didn't stay with them long, though. I discovered—very early in the game—that a two-bladed propeller has vibration problems that a prop with three blades doesn't have. This potentially destructive situation always exists with propellers that have two blades. It's always there, but most of the time it doesn't give airplanes any trouble. I mean . . . when you make a turn with a plane, how large a curve do you usually fly? A quarter mile? A half mile? That's not nearly sharp enough to cause a problem. But a windplant supported in its center on a bearing whips right around, doesn't it? There just isn't any way to make a two-bladed wheel hold up on a windplant. Sooner or later—and probably sooner—it'll snap at the hub, or one of the blades will let go.

"It doesn't matter if you have one blade or a dozen: If you design them right, you can make that wheel catch all the wind that comes through it. You can stand behind those blades and strike a match, and it'll hardly blow out. You're catching all the wind, you see, and you're slowing it down and changing its direction. One blade is just as good as four or five or more.

"The only trouble with one blade, however, is that you can't balance it . . . and two blades have the vibration problem I've mentioned. A wheel with three blades nicely solves both these problems, and you'd be foolish to add any more because every time you put on another tip, you're adding unnecessary drag. That's a waste of power.

TURNING THE WIND INTO WATTS

"There's another factor involved, too. We wanted our windplants—which had 15-foot-diameter propellers—to develop their maximum charging rates in a wind of, say, 20 MPH . . . but we didn't want their tip speeds to exceed 125 MPH. A three-bladed prop met these requirements admirably.

"Well, once we had the propeller design worked out, along about 1927, we still had two main problems: speed and pressure. If you want to get as much power as you can from a light breeze, you've got to have a propeller with a large diameter. But when you have a large diameter, you've also got something you can't control in a high wind. You need some way to regulate your propeller's speed, and you want to be able to take the pressure of the wind off your blades during a real gale.

"So I developed the fly-ball governor. I mounted weights on the hubs of our propellers so the centrifugal force of higher speeds would twist all three blades identically and change their pitch. This automatically feathered the propellers in high winds. It both slowed them down and relieved the pressure against them. We set the centrifugal controls so our blades couldn't receive more than the pressure for which they were rated. We've had winds of more than a hundred miles an hour on our plants down there at the South Pole. No problem. We've had plants scattered all over the West Indies and on the Florida Keys, and we've never had a single one go down in a severe storm or a hurricane yet.

"Anyway, after obtaining a patent on our governor, we built about 20 or 25 plants out there in Montana from 1927 to 1931. They all had our new propellers and governors on them, and we sold them to ranchers in the area. We bought our gen-erators from Robbins and Myers, and we built both 32- and 110-volt DC systems. I think we got our towers from the Challenge Windmill Company in Batavia, Illinois. The towers, you know, were actually meant for water-pumping windmills. Nobody else was making windplants. We introduced the business to North America . . . I guess to the world. A few others were playing around with ideas, but we were the first to manufacture a practical machine.

"In 1931 we sold our ranch holdings—my brother was with me at the time—and I formed a Montana corporation, sold stock, and really set up to make windplants. Later, of course, I moved the operation to Minneapolis. We spent about a year or better designing and building a big generator. There wasn't one available at that time that would produce 2,000 watts of power at our working range of 225 RPM. You couldn't buy one anywhere, so we designed and built one just for our propeller.

"Now this was quite important for a couple of reasons: Number one, there's a lot more to good propeller design than most people realize, and number two, the best propeller in the world isn't worth much if the generator it turns isn't exactly matched to the prop.

"The whole idea of high-speed propeller design is to throw the wind that hits the blades . . . and throw it out quickly. You don't want it to drag all the way along the back of the blades. That's a tre-mendous amount of friction—a tremendous force—and you want to eliminate it. Sometimes a very little change—a 64th of an inch—in the curve on the back of a propeller can affect its power output a seemingly immeasurable amount.

"Well, forty years ago I designed a special ma-chine that would let me determine just how effi-

BLOWING IN THE WIND

Recollections of a
Windpower Pioneer (continued)

cient a blade might be. I had a test stand made that extended out two feet past the end of a propeller, and at each foot along the arm we mounted a separate wind pressure gauge. We checked a lot of blades on that stand until we knew exactly how to design a propeller that was as efficient as we could make it.

"We had to balance the generator's load to match the efficiency of the propeller. If your blades work best at a certain RPM in a 7-1/2 MPH breeze, then they should turn exactly twice as fast when the wind blows 15 MPH, shouldn't they? They won't catch all of that 15 MPH wind unless they do, will they?

"OK. The trick is to design the generator so that its load increases just fast enough to allow the propeller to double its RPM as the force of the wind doubles. And that's what we did . . . right up to the top speed we wanted: 18–20 MPH.

"Now this wasn't easy, because a conventional generator doubles its output when its speed increases by only something like 25%. Obviously that wasn't a very good match for our propeller, so we tried several things until we finally came up with a special alloy for the field poles in the generator. We finally got a combination that made the load of the generator fit the output power curve of the propeller over the entire range of wind speeds up to 22, 23, or 24 miles per hour . . . where the blades were set to feather out.

"Then we also came up with our own special brushes in the generator. It's not too hard to set up a big DC generator and run it with a stationary engine, see, because you've got a fixed speed of operation, and you can adjust everything so it's working the best for that rate of output. Now I'm particularly thinking of the commutator arm and its brushes, which slide from one wound coil to another inside the generator. Every time those brushes move from coil to coil, you know, they want to throw a spark. When you break DC, you get an arc . . . and those flashes will burn little rust spots on the commutator, and then it'll just grind the brushes off in a matter of months.

"What you look for, of course, is the neutral zone . . . the one small area where your brushes will throw the least spark as they leave one coil and go to the next. This isn't too hard to find, and when you've got a fixed speed on your engine and generator, you can set everything just right to make use of it.

"A windplant isn't like that, though. It's set to kick its generator into operation at about 125 RPM, and it reaches full output—3,000 watts or whatever—up around 225 RPM. Now that's OK, but every time the RPM varies—and it can change a thousand times a day—the neutral zone shifts. No matter how you adjust your commutator, then, your windplant's brushes are going to be set to throw a much bigger spark than you'd like as they move from coil to coil during the greater part of the plant's operation.

"Everyone in the business faced this problem, of course, but none of the others ever licked it. We did. I developed a brush made up of a layer of graphite, then carbon, then graphite, then carbon. This gave us a brush with high cross-sectional resistance. The direct current would almost quit flowing before the brush made its jump from one coil to the next, and that was just what we wanted. We've had plants run 10 or 15 years on their original set of brushes. That's unusual!

"Voltage regulation is another tough situation you have to face with DC. To change the irregular

TURNING THE WIND INTO WATTS

power generated by the wind into a steady flow of current for use, you have to go through batteries. The only trouble is that you can't let your generator feed the same amount of electrical energy to the batteries all the time, or you'll burn out the storage cells. As a charge is built—as the battery becomes more nearly 'full'—you want to charge it at a slower and slower rate. That's why you always had to get up at two o'clock in the morning, or some other unhandy hour, and shut plants off to keep them from burning out their storage banks.

"We had the only windplant that didn't have this trouble because ours was the only one that was completely voltage regulated. Our control—we called it the Master Mind—inserted a resistance into the generator fields to weaken their output as the batteries filled up.

"Now that was a problem in itself because the Master Mind contained a set of points that had to open and close thousands of times a week. This meant thousands of arcs and flashes. Eventually the points would stick and make the generator begin to run like a motor as soon as the wind died down. That wasn't good, you know, because it would soon drain all the stored energy.

"We licked that one by developing what we called our 'reverse current relay'. We ran a little bit of direct current—opposite in polarity to the main flow—right back through the points to make them open with one quick flash instead of just hanging there, floating, until they'd burned themselves out. It was a little shunt circuit, actually, that opened and closed the main cutoff with one clean action just when we wanted it to.

"Our most important work was done in less than two years . . . from 1931 to 1933. By '33 or '34 we were in pretty good swing. We came up with a few improvements as we went along, of course, but after 1936 or '37 we ran for 20 years without making any basic changes in our design.

"Back then there weren't any experts on slow-speed electrical generation. There were no experts on voltage regulation, and nobody had ever heard of making an airplane-type propeller for a generator. There were no books on the subject . . . nothing to go by. I developed my own expertise. When you have a problem, you just stick with it until you find a solution. That's how I wound up with more than 25 patents. Every one of those patents represents a problem that we solved. But I wanted to manufacture the finest windplants possible. I'm kind of a freak, I guess. I want things to work forever. I've always built my plants to last a lifetime.

"I've had battles with manufacturers all my life. When I started looking for bearings to put in our windplants, I found out that the ones the companies made which were called 'permanent' would last about two years. The bearings themselves were pretty good, but the seals around the races would dry out and let the grease inside get away after a few years. What I did was take some of the bearings used in the rear axle of a car, mount them in a special compartment with a special lubricant, and then put my own seal over them. They'll last 20 years that way . . . and 20 years is closer to a lifetime than two.

"We've had plants that have run 25 years with no lubrication. I talked to a rancher out in New Mexico, and he's been using his for *over* 25. He's still using it, and he's never done much more than climb up once a year and tighten a few bolts and whatnot.

BLOWING IN THE WIND

Recollections of a Windpower Pioneer (continued)

"The brushes on most windplants, as you know, go out all the time. They don't last long at all. Well, I got letter about a year ago from a mission in Africa. The people there bought their plant in 1936, and that letter was their first order for replacement brushes. They've used the generator all that time. Same thing with our blades. Our old standby was aircraft-quality, vertical grain spruce. Sitka spruce from the West Coast. I used to go out and select the lumber personally and have carloads of it shipped back to the factory. During the war, I had a little trouble getting the quality I wanted. We rough-cut the airfoils first—from 2 X 8 planks—on a special machine. Then we put them aside in the kiln-dry rooms for several weeks to make sure they were completely set and weren't going to warp. After that we made our final cuts. We had a great big sanding machine that worked down both sides of a blade. It was set up like a planer or a duplicating lathe. You clamped your raw blade into mounts on one side, and then you ran a set of feeler rollers over a perfectly finished blade that was always mounted on the other side. This guided the application of power sanders to the unfinished airfoil: You could smooth it right down to the exact contours of the master quickly, easily, and automatically. Then we finished them with an asphalt-base aluminum paint. Propellers we built 25 or more years ago are still going strong.

"Another innovation was that we never put a brake on our plants: Our tail vane was enough. We had it hinged so we could lock it straight behind the generator or swing it away to the side. It would remain streamlined to the wind either way, of course, so when it was in the second position, it pulled the generator and propeller right around edgeways to the moving air. This took most of the wind off the blades, and they'd sit up there and just idle during violent storms.

"Most manufacturers fastened the vane straight behind the generator with springs, and you had to use a line from the ground to pull it around to the side. If that line broke during a gale, there was nothing you could do about it. The windplant would run away and tear itself all to pieces . . . unless you had a brake that you could apply. And brakes, for some other reasons, weren't a good idea either.

"We set our spring up the other way, see. It always wanted to hold the vane to the side, and you had to use a line to pull the tail straight back. This way, if the line broke, the vane would pull the propeller around and make it idle. Ours was designed to protect itself if anything went wrong.

"We tested some brakes when we were still experimenting out in Montana, and very quickly found that they're a source of trouble. The brake bands freeze up, and you have to climb the tower with a hammer and knock them loose. Besides that, it's not very smart to completely stop a windplant propeller. The ice mostly freezes on the lowest blade, and that'll wreck your plant if you turn it loose. It's much better to let the propeller swing around a little bit during a winter storm. What ice or frost it collects will be distributed evenly that way and won't give you any trouble.

"We must have built about 50 million dollars' worth of plants in 25 years. I can't remember, but I think we had 260 employees at one time. We could produce eight to ten plants a day working one shift, and during the war we ran three. We ran around the clock in Minneapolis, and I even bought another factory in Iowa and ran it for a few years. We didn't build windplants out there, but

TURNING THE WIND INTO WATTS

we manufactured similar equipment: electrical and magnetic hardware for the Army and Navy, gear that protected our ships from the Germans' magnetic mines . . . things like that.

"We used to sell everything you'd need on a ranch—fans, motors, electric irons, toasters, percolators, freezers, refrigerators, whatever—all built to run on 32-volt DC. Hamilton Beach manufactured them for me to my specifications. I even had a freezer that was so well insulated that you could unplug it, and it would keep ice cream frozen for four or five days. All this equipment could be powered by our windplants, of course.

"There'll always be a small, scattered market for individual plants—especially in the more remote areas of the world—but the Rural Electrification Administration has pretty well killed the demand for self-contained DC systems in this country. AC is just too readily available everywhere. Alternating current is all over the place, often at artificially low prices. That's a tough combination to beat, and I quit trying to fight it in the '50's . . . we closed the factory in 1956. I still feel that the individual DC plant is largely a thing of the past. If I were building windplants today, I'd go with AC. And I wouldn't concentrate on the small units . . . I'd think about larger ones that could feed directly into the power distribution grid that's already set up.

"As a matter of fact, I proposed just that idea to Congress back in 1952. The power companies, you know, already have a great number of steel towers set up to carry their transmission lines across the country. I added to this the fact that AC generators require almost no maintenance at all, and I came up with an idea: Put windplants right on top of the towers.

Marcellus Jacobs says the individual DC generating plant may be impractical today, but he suggests tying AC windpower units into existing power grids.

"Pick a stretch—I took Minneapolis to Great Falls for an example—and install a thousand AC windplants on the towers in between. It doesn't matter what the wind does, at least some of the generators will be producing all the time. Just let them feed supplemental power into the grid whenever the wind blows.

"The beautiful part of this plan is the fact that the wind blows strongest and most steadily when we need power most . . . in the winter. I've talked to men who manage the power grid, and they tell me electric heat has become so popular that they're now forced to keep thousands of dollars' worth of standby diesels on hand.

"But let's zero in on what the individual could do. He could go out and build his own windplant about the same way I put my first ones together, with materials found in junkyards and with other odds and ends. I haven't been active in the field for 15 or 18 years now—most of my patents are public property—and there's a lot of new equipment I'm not familiar with. But I'd say that some of the AC generators and the rectifiers now available should make the construction of a homebuilt windplant pretty easy."

BLOWING IN THE WIND

Make the Wind Your Waterboy

All too often, when back-to-the-landers plan for their electrical needs, they tend to overlook a few simple ways to cut their power consumption. For example, if the average winds in an area are high enough to make the use of wind power feasible, most folks think of electrical windplants. And with good reason, for a watt-producing windspinner can be a useful source of homestead energy. A homestead's water supply, however, can be pumped by a mill, not only reducing the need for electricity (possibly by 50 kwh per month), but also avoiding the cost of long, expensive stretches of wire to reach a remote pump.

Unfortunately, a new windmill—complete with tower and pumping equipment—can cost several thousand dollars depending on its size and the factory-to-farm shipping charges. For a person short on time and long on dollars, a store-bought unit might be good choice. But for the average family trying to make a go of it on a few acres, new equipment is probably out of the question. Consequently, folks who find themselves in such a dollar-depleted situation will no doubt be happy to hear that it's not difficult to locate, purchase, and repair a used windmill at a fraction of the cost of a setup that's fresh off the assembly line.

First of all, it's best to try to find a mill that's commonly known and is still manufactured. Thus you'll be assured of a ready source of parts. Aermotor, Baker, and Dempster are the three most widely recognized pumps that are still being made, though you'll undoubtedly run into some "extinct" brands such as Currie, David Bradley, Fairbanks Morse, and The Wonder. Unless you're an adept mechanic and are prepared to make parts for a machine that's no longer manufactured, steer clear of discontinued models.

One of the best techniques for finding a good used outfit is to keep your eyes peeled while traveling. Once you begin looking for them, you'll be amazed how many out-of-service windmills there actually are. Often a farmer will be happy to sell the apparatus at a bargain price, just to have the "eyesore" removed.

After a few weeks of watching, you'll start to become an ace at identifying different brands from a distance. For example, Dempster tail vanes taper to the rear, Baker tails are notched, and Aermotor tails are squared off. Engine covers are also a giveaway: An Aermoter lid looks like a peaked metal cap or helmet, while a Fairbanks Morse cover resembles a rural route mailbox. In short order you'll be able to save yourself the trouble of inspecting brands you're not interested in.

Another promising way to acquire a used windspinner is to advertise in your local paper. Many rural residents will be astonished that anyone would actually want their old windmills.

Once you've homed in on a likely prospect, you'll need to determine the plant's condition in order to establish its value. Even if you find an engine that's beyond repair, the tower on which its mounted could prove to be a bargain. Fortunately, windmills are relatively uncomplicated, and the inspection procedure is easy.

All of the currently manufactured windmills—and many that were built in bygone days—are equipped with a "pullout lever" within reach of the ground on one of the tower's legs. When engaged, the pullout lever folds up the tail vane and applies an internal expanding brake to the machine's main shaft. This system protects the mill from damage during storms and allows the pump to be shut off. So the first thing to do during an inspection is to

WIND AND WATER

If the average winds in an area are high enough, the windmill is a logical choice for those wishing to reduce their power consumption. Windspinners can produce watts and pump water. A new assembly—complete with tower and pumping equipment—can cost thousands of dollars, however, so time spent locating and repairing a used windmill might be a good investment.

Make the Wind Your Waterboy (continued)

release the pullout and see if the wheel points into the wind and spins.

Next, stick a medium-sized adjustable wrench in your back pocket, station a friend at the pullout lever in case of an emergency, and proceed up the tower to check the mill's insides. Remove the steel cover from the gearbox—being careful to keep your hands, hair, and clothes clear of the working parts—and observe the motion of the spinning assembly inside. If all the gears and the ball bearings on the eccentric shaft seem to work smoothly and quietly, you've probably found a worthwhile used mill.

The asking price for a previously owned water pumper can vary tremendously. In general, you should never pay more than half the price of a new one, even for a plant in nearly perfect condition. And remember, it's a buyer's market.

Once negotiations are complete, you'll need to figure out how to move the windmill to your own location. Small plants can be handled by two stout individuals with a pole and a winch, but larger units will probably require a crane or cherry picker (usually well worth the expense). Keep in mind that an 80-foot tower can weigh over a ton, so don't attempt more than you can handle.

Start by disconnecting the pump rod, which links the engine to the suction rod where it enters the ground. Then climb the tower, unbolt the sail, and lower it to the ground, piece by piece if necessary. If you can reach it safely, you also might remove the tail at this point. Now you can either lower the whole assembly to the ground—for towers up to about 30 feet—or disassemble the tower section by section.

Most windpowered water movers have their pumps down in the well, so you'll have to remove

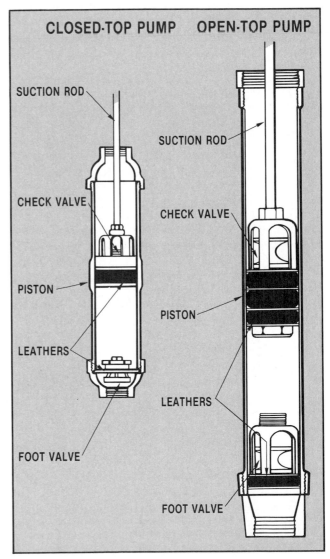

CLOSED-TOP PUMP   OPEN-TOP PUMP

SUCTION ROD

SUCTION ROD

CHECK VALVE

CHECK VALVE

PISTON

PISTON

LEATHERS

LEATHERS

FOOT VALVE

FOOT VALVE

the suction rod, pump column, and pump from the hole. Since both the rod and the column are assembled in sections that screw together, you can lift them and disconnect each piece as you reach its lower joint with the appropriate pipe wrench. Columns come in sizes between 1-1/2" and 8", and the rod is usually made of wood and has 1" to 2" metal couplers. If the mill you've selected has a very large pump column and is several hundred feet deep, you may need a crane just to lift that much pipe. Even 2-1/2" material weighs almost six pounds per foot.

The pump cylinder on a used plant will be one of two types: either open- or closed-top. The former allows the piston to be removed from the casing for maintenance by simply lifting the suction rod, while the latter requires that you pull out the column in order to get at the pump's guts. Once you've extracted the pump from the well, determine which type you have and—if it proves to be a closed-top model—decide whether you want to take the chance of having to pull all that pipe out of the ground again to replace the washers. You may choose to buy an open-top replacement.

Whichever type of pump you end up using, one that's been previously owned should be inspected and rebuilt. Replace all the washers, gaskets, cup leathers, and valve leathers. Furthermore, it's a good idea to have a machinist inspect the brass cylinder to determine whether there's enough wear to require replacement of that vital part.

Once a site's been selected for the windmill, you'll need to pour concrete footings to anchor the tower securely to the ground. In order for the engine to work without vibration and strain, the drive rod must be centered exactly over the well hole. A satisfactory technique is to dig four 18"-diameter, 5'-deep holes (spaced for the tower's four legs) and then to brace the mill above the holes with timber running under its horizontal girders. Next, predrilled sections of galvanized angle iron are bolted to the insides of the legs so that they'll extend about 2-1/2' below ground level into each hole. When you get the tower exactly centered and leveled, pour concrete (one part cement to three parts sand to four parts gravel) into each hole until its filled.

There's a good chance that your well isn't the same depth as the one over which the windmill stood before. Therefore, you'll have to measure the depth, by extending a two- or three-pound weight (such as a hammer) on a steel tape or cord having a known stretch factor, and either add or subtract materials to suit your hole. Allow five feet of clearance at the bottom of the well to keep the pump from picking up mud.

Getting the pump and its connecting links back into the ground is roughly the reverse of the operation you performed to get the pipe and rod out of the old well. However, bear in mind that the weight of the pump column and rod will increase as you add each new section. What started as a manageable load can get out of hand, so another visit from the crane might be in order. Also, remember that a closed-top pump must be attached to the first chunk of pump column and have its suction rod installed after the pump column is in place.

At the predetermined depth, stop adding column and rod, and anchor the assembly to the ground. Then connect the drive rod to the engine and suction rod, add oil to the gearbox, yank on the pullout lever, and watch her spin. It's a tremendously satisfying sight.

BLOWING IN THE WIND

A Spring That Gives the Wind a Lift

Folks attempting to set up a wind pump on a shoestring budget sometimes find that after buying a gearbox, tower, cylinder, and well pipe, there's not much cash left over for a manufactured "sucker rod". Unfortunately, the cost-cutting substitute most readily available for this vital connect-

ing rod is one-inch galvanized pipe, and this option is excessively heavy and will often overload the windmill and decrease its output.

However, in many cases it's possible to use a spring to help lift the heavier, make-do pumping rod. Such a booster can make a real difference in the amount of water a pump supplies, especially in light breezes.

So, if you find yourself struggling with a too-heavy sucker pipe, start looking for a really big spring: A garage door coil is usually ideal. Using an eyebolt, connect one end of the spring close to the top of the tower. Attach the other end to your substitute sucker rod with a tail-pipe clamp. You can also install a turnbuckle, but this is optional since the spring tension can be easily adjusted by moving the clamp up or down on the rod.

The spring should be stretched out far enough to support the weight of the sucker pipe, thereby allowing the energy your mill captures to lift only water. So, in order to balance the spring's load, disconnect the makeshift rod from the gearbox by unbolting the jackshaft. Common sense will tell you that when the rod drops down, the spring is too loose, and when the pipe rises, the tension is too great. Once no motion occurs, you'll be all set.

Remember that if your pump has the normal eight-inch stroke, you'll want a spring that—at rest—is at least double that length, so that it will exert a more even force through the stroke. Folks with exceptionally deep wells might want to use more than one spring, spacing them equally around the tower to keep the rod centered.

This is an easy modification that certainly pays off for a wind pump with a homebuilt sucker pipe. Even in a stiff breeze it can increase the system's per-minute flow by as much as 25%.

For a water-pumping windmill that's put under a strain by the weight of its "sucker rod" or the distance it has to lift the water from the ground, such a simple device as a helper spring might well be the perfect answer.

WIND AND WATER

The South Pacific Wind-washer

Here's an idea brought back by service personnel stationed in the South Pacific during World War II that might be useful to homesteaders living beyond the reach of power lines or to anyone with a real yen to conserve energy or declare independence from the utilities.

Not only was there a lack of any kind of automatic washer on those islands, but even if there had been a handy machine available, there was often no electricity, gasoline, or other power source to run it.

But those innovative folks soon found a way to take advantage of what they did have: water and nearly steady tropical winds. They rigged up—by the hundreds—simple, inexpensive, and practical washers that were directly powered by the breeze.

The actual "machinery" of one of these devices was nothing more than a four flat propeller blades mounted with wedges onto a disk that was bolted

to a pipe. This, in turn, was connected to a simple set of swivel joints. As the wind blew, the blades revolved, the pipe turned, the joints rotated, and the plunger—which was just a length of conduit or broomstick with a round board bolted onto one end—moved up and down inside a chopped-off 55-gallon drum.

Since dirt, bugs, and other debris often fell into the washers, some machines were further enhanced with lids. These covers were usually made in two pieces from plywood or sheet metal and were either just laid over the drum's opening or were hinged to the barrel's sides to facilitate loading and unloading the washer.

The uncomplicated clothes-washing operation consisted of tossing in the grimy duds, pouring in enough water to cover them, adding soap or detergent, and facing the sawhorse-mounted prop into the wind.

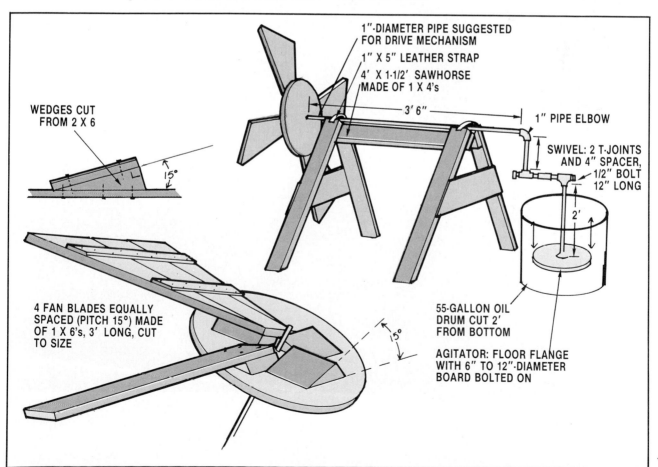

WEDGES CUT FROM 2 X 6

15°

4 FAN BLADES EQUALLY SPACED (PITCH 15°) MADE OF 1 X 6's, 3' LONG, CUT TO SIZE

15°

1"-DIAMETER PIPE SUGGESTED FOR DRIVE MECHANISM

1" X 5" LEATHER STRAP

4' X 1-1/2' SAWHORSE MADE OF 1 X 4's

3' 6"

1" PIPE ELBOW

SWIVEL: 2 T-JOINTS AND 4" SPACER, 1/2" BOLT 12" LONG

2'

55-GALLON OIL DRUM CUT 2' FROM BOTTOM

AGITATOR: FLOOR FLANGE WITH 6" TO 12"-DIAMETER BOARD BOLTED ON

HOLDING YOUR VOLTS

While electricity can be generated by capturing energy from sunlight, wind, or water, the process of converting these natural forces to volts and amps is only part of the task. Photovoltaic cells can produce current only while the sun is shining, and windplants can operate only while the air is moving past their blades. And, though water-driven generators yield a more continuous source of electricity, outputs can fluctuate, and they—like the other power-producing systems—are often unable to instantly adapt to the changing demands placed on them.

Therefore, to stabilize a home power system and to provide electricity when none is being produced, banks of batteries are needed. There are several different types of storage cells available, and when choosing, you'll have to consider cost, battery characteristics, and the demands that will be placed upon the storage bank.

HOLDING YOUR VOLTS

Charge and Discharge:
A Battery Primer

Making the switch from dependence on the commercial power grid to an independent electrical generating system is undoubtedly an admirable goal, but it's important to understand that the generation of the electricity—whether it's done with a unit powered by sun, wind, or water—is likely to be only half the job of setting up a home-scale power system.

Because the output of most small generating systems is limited in quantity or delivered only intermittently, some method of storing electricity is usually a necessity if the system is expected to supply a homestead with a sufficient amount of power to meet heavy demands and to provide a continuous source of electrical energy.

This is most often accomplished by incorporating a bank of batteries into the system, and since different batteries have different characteristics and capabilities, it's important to select the type best suited to the individual application. Because of such considerations as initial cost and the ability to be recharged a great many times, lead-acid batteries—rather than nickel-cadmium or other types of cells—are usually selected for use in homebuilt generating systems.

Batteries fall into two general categories: primary cells and secondary cells. The latter are more commonly called storage batteries. The familiar flashlight battery is an example of a primary cell and is composed of chemicals which react with one another to produce electricity. Once these chemicals are exhausted, the cell ceases to function and, despite information to the contrary, cannot be recharged. This makes primary cells unsuitable for anything other than one-time use.

While like primary cells in that they generate electricity by means of a chemical reaction, storage batteries are capable of being recharged, and electricity drawn from them can be replaced by charging them from an external generator such as a windplant, photovoltaic cells, or a hydroelectric unit.

Of course, even among secondary cell batteries there's quite an assortment. But despite recent encouraging development in battery technology (such as the refinement of the nickel-cadmium battery), lead-acid cells provide the best proven, cost-effective performance.

The brainchild of nineteenth century English chemist R.L.G. Plante, the lead-acid cell—in its simplest form—consists of two lead plates immersed in a weak sulfuric acid solution that serves as an electrolyte. When voltage is applied across the two plates, a change takes place. The plate connected to the positive side of the generator absorbs oxygen from the electrolyte and becomes lead peroxide, while the negative plate assumes a spongy appearance without changing form chemically.

When the charging current is removed, a potential difference of about two volts remains across the plates. This stored electrochemical energy can be tapped as a power source, and—once exhausted—it can be replaced. If no energy losses occurred during the process, this transformation could be repeated indefinitely, and all the power that was put into the cell could be drawn back out again. However, certain physical restraints prevent the cell from working at such efficiency.

It didn't take Plante's successors long to figure out that the use of smooth lead plates limited ily deformed during manufacture. Consequently, small amounts of antimony alloy are often added to the lead to make it more rigid.

The density and porosity of the paste used in a battery, along with the physical properties of the grid, determine the capacity and discharge rate of the cells. If a high discharge current is desired (as is the case with a car battery, for example), the plates are often a mere 1/8″ thick and have very porous paste.

Such thin plates, however, can't tolerate many deep discharge cycles. Because the positive plate reacts with the electrolyte, part of the lead paste is shed from the grid every time a cell is discharged. The number of charge/discharge cycles that a battery can stand is determined both by the thickness of its plates and by the depth of each typical discharge.

Therefore, in order to build a battery that can withstand many deep discharges, the plates must be made thicker and the paste denser than are those in a rapid-discharge battery. Of course, if an instantaneous high output of energy is required, a number of these deep-cycle units will be needed ... since their limited plate-surface area prevents each one of them from giving up great amounts of power quickly.

Choosing between discharge rate and life cycle is only one compromise that you'll have to make when selecting a battery. For instance, the antimony alloy that makes plate construction easier presents some problems of its own. As the cell cycles, the antimony will be leached from the positive plate, and this can lead to two problems.

First, the antimony has a nasty habit of crossing to the negative plate and contaminating the

the rate at which acid could react with the metal. Consequently, plates made of a porous lead paste, which allowed the acid to contact a greater surface area, were soon developed. They were effective enough to allow the size of the cells to be decreased significantly without any loss of power storage capability.

The lead paste alone isn't rigid enough to hold the plates together, though, and the material must be supported by a wire mesh (called a grid), which is also made of lead. This system presents production problems, since the soft metal is eas-

A bank of batteries is needed to store the electricity produced by a home generating plant. Such a setup might well consist of many 2-volt, pure lead cells tied together to meet the current and voltage requirements of the system.

HOLDING YOUR VOLTS

Charge and Discharge:
A Battery Primer (continued)

surface, reducing the capacity of the cell. Furthermore, the roving metal ions also conduct some current from plate to plate. This exchange is entirely independent of any external demand, and it can slowly discharge a battery's cells even though the unit sits unused. In fact, a substantial portion of a charge can be lost in as short a time as 90 days.

To combat self-discharge problems, scientists developed a new method of hardening the lead grids. By adding an alloy with less than 0.1% calcium, they managed to improve plate rigidity without affecting the cell's chemistry. As a result, internal currents were cut to less than 1/10 of what they were before. However, the calcium-reinforced grid demands its own trade-offs. Such batteries can be used only in shallow-discharge situations and should normally be cycled within the upper 20% of their rated capacity. Furthermore, the stress of deep discharge tends to expand the calcium grid, often causing the paste to lose adhesion.

Despite all the developments in alloy-augmented-grid batteries, however, the basic lead plate that Plante used still offers the longest life and most reliable service. What's more, advances in plate construction (as well as in manufacturing procedures for pure-lead-grid and -paste batteries) have produced units of outstanding performance. Unfortunately, their initial cost, which is approximately twice that of the alloyed variety, limits their application.

When assembling a battery bank, a person first needs to know the capacity that will be required. This involves determining both your anticipated load and the capabilities of the batteries you intend to use.

The condition of a battery can be easily and quickly determined by measuring the specific gravity of the electrolyte with a hydrometer. However, the measurement will vary with the temperature and should be corrected by adding or subtracting the number listed in the diagram opposite the temperature readings.

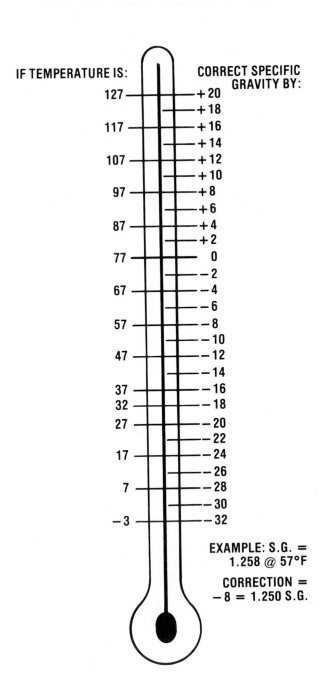

SPECIFIC GRAVITY CORRECTION FACTOR

IF TEMPERATURE IS:

CORRECT SPECIFIC GRAVITY BY:

IF TEMPERATURE IS:	CORRECT SPECIFIC GRAVITY BY:
127	+ 20
	+ 18
117	+ 16
	+ 14
107	+ 12
	+ 10
97	+ 8
	+ 6
87	+ 4
	+ 2
77	0
	− 2
67	− 4
	− 6
57	− 8
	− 10
47	− 12
	− 14
37	− 16
32	− 18
27	− 20
	− 22
17	− 24
	− 26
7	− 28
	− 30
−3	− 32

EXAMPLE: S.G. = 1.258 @ 57°F

CORRECTION = − 8 = 1.250 S.G.

Although many people feel more comfortable discussing wattage (and, in fact, many power sources are rated that way), storage batteries are rated in ampere-hours. (An amp-hour is equal to one amp of current flowing for one hour.) You multiply the current by the elapsed time to arrive at the amp-hour rating. Thus, ten amps flowing for one hour produce the same amp-hour rating as does one amp flowing for ten hours.

To illustrate the method used to determine the storage capacity required of a system, consider the example of a hypothetical house that's equipped with a small 12-volt windplant. First, assume that the residents of the home use about 100 amp-hours a day.

Next, say that the turbine gets enough wind to charge the batteries three times per week. Since the recharge interval will be about two and a half days, 250 amp-hours of storage will be needed to satisfy the home's requirements. But suppose the wind didn't blow for five days? After two-and-a-half days the battery bank would be exhausted. To handle such occasional doldrums, the storage system would need at least 500 amp-hours of capacity. And such a larger system would allow the batteries to cycle in the upper half of their range most of the time, improving the life expectancy of the batteries.

The battery setup in this example would be considered a deep-cycle system. And the choice of battery for a given application is—as already hinted—largely determined by the depth of discharge to which it will be subjected. Since battery banks can be expensive, and there isn't always a distinct line between deep and shallow cycling, examine the operating conditions of each classification carefully.

A shallow-cycle environment is usually specified as one in which the batteries are seldom if ever drawn down to less than 80% of full charge. Such a situation occurs more often with photovoltaic and hydropower sources than with windpower plants. (Batteries are often used with hydroelectric systems to supplement the generators during peak demand periods, thus allowing a smaller, less expensive generator to be used.)

Almost any battery will work in a shallow-cycle installation, but some are better suited to it than others. For example, even though antimony-grid batteries are the least expensive to purchase, they have some serious drawbacks for shallow-cycle applications.

First, an antimony cell has a significantly shorter life span than either its calcium or pure-lead counterparts. What's more, the antimony battery is less efficient. In fact, for every 100 amp-hours obtained from an antimony battery, about 125 will have to be replaced. By comparison, a calcium battery can yield 100 amp-hours for an investment of only 110.

In a shallow-cycle system, the heartiness of a pure-lead cell wouldn't be put to full use, and its minor advantage in life span over the antimony-grid type is clearly outweighed by its much greater cost. Therefore, the calcium-grid type seems to be the logical choice.

A deep-cycle battery bank, on the other hand, is often discharged to as little as 20% of its total capacity, thereby placing a great amount of stress on the cell plates. Of course, the description "deep cycle" is entirely relative, and with the addition of enough batteries, any deep-cycle setup can be altered to function in a shallow-cycle mode.

Charge and Discharge:
A Battery Primer (continued)

Because calcium-grid batteries won't hold up in a deep-cycle situation, look at only antimony and pure-lead cells for such applications. The particular type—as well as the grid construction and paste density—that'll prove to be most suitable will depend largely on the rate of discharge (or charge) of the batteries.

To give you an idea of how important it is to match a battery's performance characteristics to the load, look at how manufacturers rate batteries and see what might happen if you violate the cell's use profile. The industry standard for current rating is a measure of the amount that can be removed, by a total discharge, in eight hours. For example, if a battery is rated at 100 amp-hours, it can give up 12-1/2 amps per hour for eight hours.

But suppose you need to extract 20 amps of current per hour from that battery. Under such conditions the battery's rated capacity changes: It will be totally discharged in just four hours and will supply only 80 amp-hours instead of the rated 100. In fact, if you try to remove 45 amps, the battery will be flat in only one hour.

And what happens to the rest of the power in such a situation? Actually, it remains inside the cells. You just can't get at it . . . for the moment. When a battery is forced to discharge more rapidly than it's designed to, the electrolyte becomes stratified with the weakest portion of the acidic solution lying next to the plates, and the chemical reaction is slowed. If, however, the cells are allowed to rest (which gives the electrolyte a chance to stabilize), the rest of the charge can be removed.

On the other hand, when a charge is pulled from a battery at a slower rate than the manufacturer specifies, the cells will be able to yield more than the advertised amount of energy. A 100-amp-hour battery that's discharged at a rate of five amps per hour, for example, will yield a total of 120 amp-hours over a 24-hour period. But there's still a price to pay. To get the extra capacity, at least 120 amp-hours must first have been put in, at a very low charge rate.

Also, just as a too rapidly discharged battery can't yield its full rating, one that's charged too quickly won't reach its full capacity, even though the cells appear to be full. If the "abused" battery is allowed to sit for a few hours, the charged voltage will gradually drop. Eventually the electrolyte will stabilize, and the cell can be brought up to, and even beyond, its rated capacity with a gradual "trickle charge".

Between the two extremes of shallow and deep cycling lies a great middle ground where it's often difficult to make decisions about which type of battery would be best. What happens, for example, when the discharge varies between 30% and 40%?

Calcium-grid batteries are eliminated, since they won't last long when drawn below 20% of their capacity. But to choose between antimony and pure-lead types, a few more things about the energy system in question need to be known. For one thing, the rate at which the batteries will be charged must be determined. Wind generators, for example, often apply heavy charge rates to batteries, though the discharge rate might be, for the most part, gradual.

In such a situation, either antimony or pure-lead batteries with fairly porous plates (which allow high charge or discharge) will work. The major question to be answered before making a decision, then, is whether the added efficiency and the slightly better life span (in this situation) of a pure-lead unit would justify paying twice the price of the antimony battery.

Since the current produced by a cell is the result of a chemical reaction, temperature can play an important role in performance. Batteries are usually officially rated at 77°F, and as the temperature falls, capacity declines. A battery's power will drop almost to half at 10°F. Therefore, at such a temperature, two batteries would be needed to do the work that one could handle at 77°F. In fact, if the cells happen to be of the calcium-grid type, the decline in performance could be even greater, since those units are particularly sluggish in cold weather.

Fortunately, the sulfuric acid in the cells acts as antifreeze and prevents lead-acid batteries from freezing at most commonly incurred temperatures. A cell that's completely charged won't freeze until the mercury drops below –55°F! At less than full charge, however, perfectly good batteries can be ruined by a severe cold spell.

Conversely, cell activity increases when the temperature rises above 77°F, though this boost in performance is accompanied by a reduction in battery life. At 95°F, a battery may achieve only 50% of its normal life expectancy before the acid (which becomes more active as heat rises) corrodes the plates beyond use. Furthermore, internal currents—the ones that produce self-discharge—double with every 15°F increase in temperature.

The best approach, then, is to insulate your battery bank against temperature extremes. In addition, the storage area should be well ventilated to prevent the accumulation of the toxic and/or explosive gases that the batteries produce. Insulation and ventilation are somewhat at odds, but in most climates a small building with wall and roof insulation and a pair of vents (one high on one wall and the other low on the opposite side) should keep a battery bank comfortable and safe.

A lead-acid battery will provide years of reliable service if it's properly used and maintained. To help make sure you'll get the best return from your investment, always observe these rules:

[1] Choose the right type of cell for the application. Don't expect a calcium-grid battery to do a job better left to an antimony-grid unit, or vice versa.

[2] Have sufficient storage capacity to insure that the batteries cycle within their specified range.

[3] Use the appropriate regulator to control charge level. Don't use a wind-charger regulator with a photovoltaic system, for example.

[4] Make a habit of measuring the specific gravity of the electrolyte periodically. A fully charged cell, at 77°F, should measure about 1.250. (Use the correction diagram on page 142 to adjust for temperature.)

[5] Always keep an eye on the water levels, and add water as necessary.

[6] Keep the post connections tight and clean. Wipe off any spilled acid immediately, and neutralize the case with a solution of baking soda in water.

With the right kind and number of batteries to suit an alternative energy system, a properly operating battery bank is the perfect "middleman" between the source of energy production and the consumers of that power.

HOLDING YOUR VOLTS

Checking Your
Charge Account

In alternative generating systems incorporating banks of storage batteries, it's important to be able to monitor the condition of the power-holding cells and the amount of charge they contain at any given time. Monitoring can be done with a hydrometer, but this method is time-consuming and messy. An electronic monitoring system that keeps a continuous eye on the bank and announces when the charge is too high or too low is certainly desirable.

As a battery's charge drops, so does the voltage across its poles, and measuring that voltage can provide an accurate gauge of a battery's condition. The battery monitor shown here is essentially an easy-to-read voltmeter, on which the amount of charge in the battery being tested is indicated by one of the ten LED's (light-emitting diodes). The scale of the readout is arranged so that each of the consecutive lamps represents a 1/2 volt difference in the charge, with the first indicating 10.5 volts, the second 11.0, the third 11.5, and so on up to 15.0 volts.

The lights are switched on and off by an integrated circuit (designated LM3914), which is actually a large collection of transistors in one package. In addition, the lights are color coded to indicate the urgency of the situation. The lamps indicating 12.0 through 14.5 volts are green signifying there is no problem. The 11.0 and 11.5 lights are yellow and mean the charge is too low, and the 10.5 light is red, indicating that the battery is almost entirely discharged.

At the other end of the scale, the 15.0-volt indicator is also red, warning of excessive charge voltage which can cause a battery to "boil". This situation might occur if the temperature control in a voltage regulator goes haywire.

Finally, to be sure you'll be warned of a potential disaster, the lamps with the lowest and the highest readings are hooked to a piercing alarm. Whenever voltage drops too low or soars too high, a second integrated circuit (known as 4011 CMOS) will sound the call for help.

This monitor has been designed for easy assembly, and all the parts are readily available at electronics stores. Though most of the components can simply be inserted into the circuit board and soldered in place, a few must be carefully positioned. Each LED, for example, has a positive and a negative side, and if it's not properly oriented, it won't light. Furthermore, soldering the LED's on the foil side of the board will make it easier to mount the board later.

Integrated circuits can also be accidentally reversed, so be sure to observe the locating dot that's found on one end. In addition, the zener diode and the 220-microfarad capacitor must be positioned for proper polarity.

As a final note on the construction process, be sure to use a low-wattage soldering iron and rosin-core solder. While the components can be housed in any one of several fashions, a general-purpose utility box will do the job quite well.

A voltmeter can be used to calibrate the battery monitor. Do this by connecting the two sensor leads from the monitor to the appropriate battery terminals, clip on the voltmeter wires, and adjust the 50k-ohm potentiometer (VR1) until an LED corresponding to the voltage indicated on the meter lights up.

If you don't have (or can't borrow) a voltmeter, don't fret. Just turn the 50k-ohm potentiometer all the way clockwise, and your calibration should be reasonably close.

This neat little battery monitor will keep an eye on the condition of your battery storage bank.

Batteries should be measured when under load, since open-circuit voltage isn't an accurate indicator of battery condition. (Only a small amount of current—an amp or so—is needed.)

Also, the leads need to be connected as close to the terminals as possible, because when power is demanded from the battery, the leads themselves can produce a voltage drop. Also keep in mind that a large number of battery problems are caused by loose or dirty connections. In fact, if the audible alarm goes off during a heavy discharge session, there's a good chance that the connections, rather than the batteries, are at fault.

Even with a continuous electronic monitor, the electrolyte level in the batteries must be inspected on a regular basis. Add distilled water, if possible, though potable water will do if the demineralized liquid is unavailable. (If you wouldn't drink it, don't put it in your batteries.) And, even though your battery monitor should keep you well informed of your storage system's health, it's a good idea to check the bank with a hydrometer about once a month.

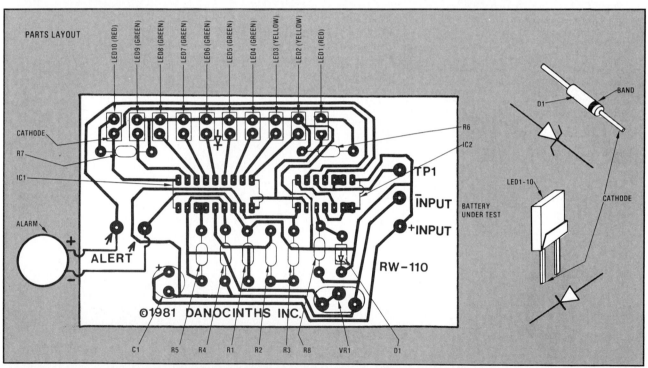

LIST OF MATERIALS

Circuit Number	Description		
R1	10K-ohm, 1/2-watt, 10% resistor	D1	6.2-volt zener diode
R2,3,4	1.5k-ohm, 1/2-watt, 10% resistor	LED1,10	red light-emitting diode
R5,8	1k-ohm, 1/2-watt, 10% resistor	LED2,3	yellow light-emitting diode
R6,7	680-ohm, 1/2-watt, 10% resistor	LED4,5,6,7,8,9	green light-emitting diode
C1	220-microfard, 16-volt capacitor	ALARM	do not substitute
1C1	LM3914 integrated circuit	VR1	50k-ohm potentiometer
1C2	4011 CMOS integrated circuit	Printed Circuit Board	

THE METHANE MIRACLE

At about the same time the world began to recognize the growing shortages of the commonly used nonrenewable sources of energy (and the attendant rises in the costs of oil, coal, and natural gas), serious consideration was also being given both to waste disposal and pollution, and to the increasing need for fertile topsoil on which to grow food to feed the world's increasing population.

Interestingly enough, at least a partial solution to all three of these problems can be found in a process in which vegetable, animal, and human wastes are used to generate a methane-rich gas. When the organic "garbage" materials in a biogas generator have produced all the burnable gas of which they are capable, the sludge that remains makes an excellent organic fertilizer.

The gas that's produced is similar to natural gas and can be used for many of the same functions: fueling gas lamps, cooking, and heating. It can also be fed into internal combustion engines, which—in turn—can be used to power generators or equipment. And the digesters used to manufacture biogas are practical on scales ranging from a small unit serving a single household to a larger system which would be co-owned by a small group of neighbors.

THE METHANE MIRACLE

Doing It Big in Biogas

The process for—and the advantages of—extracting valuable fuel and rich, life-yielding fertilizer from matter that otherwise would create an expensive disposal problem was outlined by Ram Bux Singh in a book originally published by THE MOTHER EARTH NEWS® under the title, *Bio-Gas Plant: Generating Methane From Organic Wastes & Designs With Specifications.* The information contained here has been excerpted from that book.

A fact not generally known is that manure is capable of yielding a gas very similar to natural gas in its makeup. The fuel that could be recovered from two billion tons of manure would total about 24 billion therms.

How can this gas be recovered? By using a biogas plant, in which dung and other animal and vegetable wastes can be fermented anaerobically and can be made to release a gas which is mostly methane. In addition, this process will yield a fertilizer that's rich in both nutrients and humus.

The biogas plant does essentially what composting does, but goes about it in a different way, which results in a different end product. Both are processes of fermentation: Composting is accomplished right in the atmosphere and is therefore called aerobic, while the biogas plant keeps the fermenting material isolated from the influence of air and is thus anaerobic.

Aerobic fermentation produces quantities of the noncombustible gas carbon dioxide (CO_2)—which escapes into the atmosphere—as well as large amounts of heat (temperatures can spontaneously reach 120°F). This heat does kill disease germs in the compost heap, but it is otherwise seldom taken advantage of and usually dissipates into the surrounding air. On the other hand, anaerobic fermentation results in the production of *some* CO_2, but produces a great quantity of methane (CH_4), as well. About 60% of the gas is methane, which is the same combustible material that makes up refined natural gas. The gas produced in a biogas plant does not escape into the atmosphere. Rather, it's captured by the same apparatus which serves to keep the air out. No great amount of heat is produced to be wasted, as is the case with aerobic fermentation, but the disease germs in the raw material are effectively killed by the lack of oxygen.

The process of anaerobic fermentation is the result of the digestion of raw waste material by various bacteria. The simple molecules released by the bacteria going about their life processes ultimately lead to the production of methane. The digestion is carried out in stages as different types of bacteria become active. First, the complex organic molecules in the raw material are broken down by acid-producing bacteria into simpler compounds such as sugars, alcohols, glycerols, peptides, and amino acids. When these substances accumulate in sufficient quantity, a second group of bacteria which manufactures methane can be supported. The bacteria—and especially the latter group—are quite sensitive to the conditions in which they must function. Temperature, acidity, and the concentration of certain elements such as carbon, nitrogen, and oxygen will each affect the speed and quality of the digestion.

Overall digestion by the bacteria *can* occur in temperatures ranging from 32°F to 156°F, although the production rate is faster at the higher temperatures. In turn, the production of gas will

APPLICATION	SPECIFICATION	GAS CONSUMPTION (CU. FT./HOUR)
Gas cooking	2"-dia. burner	11.5
	4"-dia. burner	16.5
	6"-dia. burner	22.5
Gas lighting	1-mantle lamp	2.5 to 3
	2-mantle lamp	5
	3-mantle lamp	6
Refrigerator	18" X 18" X 12"	2.5
Incubator	18" X 18" X 18"	1.5 to 2
Boiling water		10/gallon
Running engines	Converted diesel or gasoline	16–18/horsepower/hour

IT'S GOOD WHEN IT'S ROTTEN

decrease rapidly if the temperature is less than 60°F and will slow to almost a stop at 50°F, though the digestion of the material continues to some extent. The gas-producing reactions proceed best in two ranges of temperature, 85°F to 105°F and 120°F to 140°F, with different types of bacteria thriving in the two ranges. Unfortunately, those active in the higher range are more easily affected by environmental changes, making fermentation in this range less stable.

An excessively acid solution can also hamper the fermentation process. The best range for rapid fermentation is pH 6.8 to 8.0. If the pH goes too high, the acidic carbon dioxide formed by the digestion process will bring it down, but if it should be too low, the gas-producing bacteria will be unable to use up the acids quickly enough, and digestion will stop. It will be some time before the balance is restored and digestion can resume. Introducing fresh raw material at too high a rate is one factor that may cause the fermenting solution to become too acidic.

The amount of carbon and nitrogen in the material to be digested also affects the nature of the process. The bacteria require both elements in order to live, but they use up carbon about 30 to 35 times faster than they use up nitrogen. If the ratio of carbon to nitrogen in the raw material is about 30 to 1, then the digestion process will proceed at the optimum rate . . . other conditions being favorable. If the ratio is higher, the nitrogen will be exhausted while there is still a supply of carbon left. Some of the bacteria will then die, and the nitrogen in their cells will become available to the remaining bacteria. In the process of taking up that nitrogen, some carbon will be oxidized to carbon dioxide and will escape from the material. Thus the concentration of carbon in the fermenting material will be reduced, while the nitrogen available to the bacteria will be renewed. At that point the digestion can continue, but the process will be much slower than it would have been in material with a better ratio.

If the ratio is too low, on the other hand, the carbon will be exhausted before the nitrogen. When the carbon is exhausted, no further fermentation will take place, and the remaining nitrogen may be lost from the material. This lowers both the fertility of the digested matter and its value for agricultural applications. The goal of anaerobic fermentation in the biogas plant is to convert all the available carbon to methane and carbon dioxide with as little loss of available nitrogen as possible.

The role of oxygen in anaerobic fermentation should be obvious. It can only hinder the process by changing the digestion to an aerobic operation. As aerobic fermentation produces no methane, any oxygen unwittingly introduced into the fermenting material will cause the methane production to fall. Methanogenic bacteria cannot survive in the presence of oxygen . . . but more important, *you* may not be able to either. Biogas—like most gaseous combustibles—is extremely explosive in the presence of small amounts of oxygen.

The anaerobic fermentation of organic matter proceeds best if the slurry contains from 7% to 9% solids. The usual raw materials to be fermented in a biogas plant contain at least 18% solids, so they're often diluted with water to help the process.

The biogas plant must be constructed and operated so that it will provide the environmental

ANALYSIS OF SLURRY OF BIOGAS PLANT 50% cow dung and 50% other organic wastes used for digestion on dry weight basis			
PARTICULARS	PRIMARY DIGESTER	SECONDARY DIGESTER	CONDENSATE WATER FROM PIPELINE
Percentage of Moisture	82.2	78.6	
Volatile Organic	62.5	60.8	
Percentage of Total Nitrogen (dry matter)	1.38	1.26	pH = 7.4 Sediment 4.2 g/liter
Percentage of Phosphorus (dry matter)	1.01	1.00	
Percentage of Potassium (dry matter)	0.63	0.61	

THE METHANE MIRACLE

Doing It Big in Biogas (continued)

conditions enumerated here. Central to the operation and common to all plant designs is a digester: an enclosed tank, which is sealed off from the atmosphere with some means for filling and emptying *and* for drawing off the gas produced. Differences in design are based on the material to be fed to the digester, the cycle of fermentation desired, and the conditions under which the plant will operate.

Plants designed to digest waste matter in liquid or suspended solid form can use pipes and pumps for the feeding and emptying operations. Animal wastes, such as dung, can be processed in these plants. With other designs, the entire circulation can be accomplished without pumps. In such a system, fresh slurry can be fed by gravity into the tank from a high level . . . which allows the old slurry to overflow from the tank because of displacement. With this method, small bits of chopped vegetable matter can be introduced into the fermentation tank. Such solid matter would quickly clog most pump systems. Vegetable waste (and a gas plant capable of accommodating it) may be desirable if the animal waste available has too low a carbon/nitrogen ratio or if large amounts of gas are required. The cellulose from the plant material provides more carbon than the nitrogenous animal waste.

Plants built for a continual cycle are usually operated on a definite schedule. The period of time required for complete or nearly complete digestion of animal wastes, such as dung, is 50 days at a moderately warm temperature. A batch of such matter, if allowed to ferment undisturbed for the full period, will produce more than a third of its total gas in the first week, more than a quarter of its total in the second week, and corre-

spondingly less in succeeding weeks. To prevent such unevenness, small amounts are fed into the digesting tank daily or at other intervals . . . the amount and the interval being calculated so that the entire contents of the tank will have been replaced at the end of a predetermined length of time. By using such a technique, there will always be some fresh slurry in the tank producing gas at a good rate.

The period for replacing the contents is sometimes set at a bit less than the full digestion period . . . thus reducing the amount of nitrogen converted to ammonia. One situation which calls for such a shortened cycle develops if the slurry leaving the gas plant is not to be spread on the fields immediately, but dried in storage areas first. The ammonia, which is in solution, would be lost to the atmosphere as the water evaporates out of the slurry. However, if the manure will be applied to the field immediately upon leaving the plant, the slurry should be left in the digester for close to its full digestion period. Such a procedure dictates feeding less fresh dung per day than would be added otherwise.

The fermentation period for plant waste is somewhat longer than that for animal waste, requiring perhaps 60 to 70 days for the total process. But pound for pound, vegetable waste results in the production of seven times more gas than animal waste, so proportionally less has to be fed in combination with the dung to maintain equal gas production. However, when vegetable waste is added, the discharged slurry may contain partially undigested matter which must be further fermented before it can be applied to the fields. In order to accomplish complete fermentation, the effluent slurry can be fed into a sec-

IT'S GOOD WHEN IT'S ROTTEN

ond and smaller digester for more thorough decomposition. (This design can also be used with a digester for animal waste only.) When digesting is done in two stages, large amounts of fresh dung can be poured into a digesting tank for the bulk of the production of gas (80% of the gas will be collected from the primary digester), and the fermentation process can be completed in the second tank ... rendering the manure safe for immediate application to the fields.

Biogas plants can also be designed to ferment heavy vegetable wastes, such as corn or sugar cane husks, *alone*. Since fermenting slurries consisting mainly of vegetable wastes do not flow through pipes easily, as solid suspensions such as dung do, vegetable digesters are not run on a continual feeding cycle. Rather, they operate on a *batch* feeding cycle, in which the tank is opened completely, filled full with material, and then sealed off from the atmosphere. Depending on the composition of the fermenting material and the temperature, gas production will begin after two to four weeks, gradually increase to maximum, and then fall off after a period of about three months. When the mixture is fully digested, the tank is completely emptied and recharged for another four-month cycle. If this system is used, it is better to employ two or more digesters in combination ... so that one will be producing gas while the other is not. Because the carbon to nitrogen ratio of some vegetable wastes is much higher than that of animal wastes, some nitrogen must often be added to tanks built for vegetable fermentation. This material can be in an inorganic chemical form, but organic waste is better.

Some means of agitating the slurry inside the digester is always desirable, though not always absolutely essential. The stirring helps keep the mixture homogeneous and prevents the formation of any scum or hard spots ... which would hinder the release of gas from the slurry. In purely animal waste digesters, the solids are in suspension ... and are thus in intimate contact with the water medium in which the methane bacteria move. Furthermore, the continual feeding of fresh waste into the tank always induces some agitation of the mass of material inside the tank, helping to expose fresh undigested matter to the bacteria. In vegetable fermenters, however, mechanical agitation is essential. Since the mixture is heterogeneous, the heavier leaves tend to settle on the bottom of the tank, and the supernatant water, which contains some bacteria, floats on top. As no fresh material is fed into such a fermenting tank for a matter of months, there is no stirring and mixing effect such as there is in continually fed systems. This greatly slows the fermentation process. Externally applied agitation—mixing the bacteria with the leaves and other materials—will speed it up significantly.

In some climates, low temperatures may be a problem. At even moderate temperatures, gas production will be decreased, and at freezing temperatures the entire digestion process is arrested. Coils can be built into the digester, through which hot water is circulated to keep the fermenting mixture at a temperature favorable to decomposition. The water supplied *may* be heated directly by burning a portion of the gas produced by the plant, but a more efficient way to heat the water is feasible if the gas is being used to run an engine. Heat from either the cylinder head or the exhaust pipe of the engine can

QUALITY OF GAS PRODUCED FROM DIFFERENT DRY MATERIALS			
MATERIAL	AMOUNT OF GAS (IN CU. FT.) PER LB. OF DRY MATTER	PERCENTAGE OF COMPOSITION	
		CH_4 METHANE	H_2 HYDROGEN	CO_2 CARBON DIOXIDE
Dry leaf powder	7.2	44.44	10.78	44.76
Sugar cane thresh	12.0	45.44	10.23	44.33
Maize straw	13.0	45.94	10.26	43.83
Activated sludge	10.0	43.82	11.82	45.90
Straw powder	15.0	46.42	9.88	44.70

THE METHANE MIRACLE

Doing It Big in Biogas (continued)

be transferred through a simple heat exchanger to the digester heating coils, and the water can be pumped through the coils by the same engine.

Gas is collected in most plants by means of a metal drum inverted over the surface of the fermenting slurry. This drum is free to rise and fall inside the tank as gas accumulates and is withdrawn. The sides of the drum are inside the slurry, which seals it from the air and prevents the gas from escaping. The weight of the drum provides the pressure which forces the gas out of the tank (through a small valve in the top of the drum) to its point of use. Excess pressure will slightly decrease gas production, but the drums on smaller gas plants are not so heavy that they will have to be counterweighted. (On a five-foot-diameter plant, for example, the surface area of the slurry is about 2,827 square inches. A gas holder for this size plant might weigh about 550 pounds, which means it exerts only about 0.2 pounds per square inch of pressure on the slurry surface.)

The gas which comes out of the biogas plant is quite similar to natural gas. It usually contains about 55%–65% methane, 30%–35% carbon dioxide, some hydrogen, some nitrogen, and sometimes traces of hydrogen sulfide or other gases. The gas has a calorific value of no more than 600 British thermal units (Btu) per cubic foot. Natural gas contains closer to 80% methane, which gives it a 1,000-Btu-per-cubic-foot rating. Biogas, when treated to remove the carbon dioxide, will come much closer to this figure.

The relatively high percentage of carbon dioxide in biogas contributes nothing to combustion. The gas can be cleansed of carbon dioxide by passing it through limewater before use, in order to raise the heat content. However, it will not usually be necessary to do this when the fuel is used on properly designed burners or properly converted engines.

If the biogas has any significant concentration of hydrogen sulfide, though, it will have corrosive tendencies. The contaminant can be removed by passing the gas through a box filled with iron filings as it leaves the plant. If the iron filings become saturated, exposure to air will restore their absorbency. Another potential problem is that water dissolved in the gas may condense in the gas lines, collecting in low points and reducing or blocking the passage of gas. Furthermore, it may cause rust at its point of use or when in storage. The gas is passed through a box containing calcium chloride layers to remove moisture.

Biogas cannot be burned on a conventional burner with maximum efficiency. Its flame speed factor, which is a measure of the velocity at which a flame will travel along a column of the gas, is low compared with that of natural gas. This means that when biogas is fed to the burner built for natural gas, the flame will tend to lift off from the burner. Biogas fed at a lower pressure will stay on the burner, but may not burn efficiently . . . resulting in less heat recovered from each cubic foot of gas. The Watson House Laboratory has recommended a burner with a flame port area (the sum of the areas of the individual flame ports) to injector area ratio of about 300 to 1. Using a burner with 36 ports of 0.114 inch in diameter each, injectors with orifice diameters of 0.038 and 0.041 inch respectively, and supplying the gas at pressures ranging from 1 to 8 inches

ENGINE	BRAKE HORSE-POWER	SUPPLY LINE DIAMETER	PRESSURE IN WATER COL. IN.	GAS CONSUMPTION IN CU. FT. PER HR. (FULL LOAD)	AIR IN CU. FT. PER HR. (FULL LOAD)	WORKING EFFICIENCY LIQUID FUEL	WORKING EFFICIENCY BIOGAS
Stuart coupled with 250W, 110V DC generator	1	3/8″	2-4″	18.5	108	250W	225W
Onan coupled with 1-KVA, 110V AC generator	3	3/8″	2-4″	58.0	318	1,000W	850W
Kubota	5	1/2″	1-3″	95.0	——	5 HP	4.03 HP
Kubota	10	3/4″	1-3″	175.0	——	10 HP	8.20 HP

BIOGAS CONSUMPTION FOR DIFFERENT ENGINES

IT'S GOOD WHEN IT'S ROTTEN

on a water gauge (0.04 to 0.29 pound per square inch), they have obtained efficient, stable flames. The heat input under these conditions ranged from 3,360 to 11,000 Btu per hour per square inch of flame port area.

As a substitute for gasoline, aerated biogas may be fed into the air inlet of an engine by the following method: Gas is regulated by a gas cock, while air is controlled by a butterfly valve. The gas and air are thoroughly mixed by some material, such as copper wool, which is packed inside the carburetor. The engine should be started on gasoline, and then switched over to biogas when operating temperature is reached. Gas pressure should be regulated by means of a gasbag or some other regulating device to insure the supply at the proper pressure, and the air supply should be adjusted so that the best gas/air mixture (in the range of 6%–15% concentration) is obtained.

For use in diesel engines, aerated biogas is also fed through the air intake. Unlike the spontaneous combustion employed in a standard diesel, the compression of the biogas/air mixture is not sufficient to produce the temperature required for ignition, so there is no danger of pre-ignition. Ignition is attained by injecting a small amount of diesel fuel into the cylinder at the normal ignition time for that engine. Twenty percent of the diesel fuel normally required to run the engine will suffice for this purpose.

The slurry recovered from biogas digesters is rich in nutrients. And since the fermentation process is carried out in sealed tanks, there will be no loss of these nutrients while the slurry is inside the tank. However, when the effluent leaves the tank, nitrogen—in the form of ammonia in solution with the water in the slurry—can be lost if the liquid is stored in the open. For this reason, it is best to apply fully digested slurry to the fields as it leaves the plant. However, incompletely digested effluent from a small plant may cause harm to the fields if it is applied immediately. This is because it still contains carbon which will be metabolized by bacteria, absorbing nitrogen in the process. In the field, these bacteria will compete with the crops for their own nitrogen, often to the crops' detriment. Such slurry must be stored in a separate container until aerobic bacteria can complete the digestion.

In any case, it may not be convenient to spread the fertilizer on the fields every day, and some method of storing the slurry will have to be worked out. Closed tanks can be built to accommodate the slurry accumulation over a short period of time. Otherwise, the slurry must be dried.

The nitrogen content (N) of the effluent will range from about 1.8% to 2.4%, phosphorus (P_2O_5) from 1.0% to 1.2%, and potassium (K_2O) from 0.6% to 0.8% of the dry weight of the slurry. Unfortunately, 10% to 15% of the nitrogen may be in the form of ammonia, and if the slurry is dried, that amount will be lost. Still, this mixture has a nutrient value superior to farmyard manure. The organic matter may be well over 50% of the total solids, which makes it an excellent physical conditioner for either sandy or clayey soils.

A properly operated biogas plant, and one tailored to the needs of a given farmer's circumstances, should provide energy support along with effective waste disposal and a high quality of fertilizer. Furthermore, a well-maintained plant can offer a reliable service life of about 25 years, while being impervious to the weather.

THE METHANE MIRACLE

A Modest Methane Experiment

Perhaps you're just waiting for sufficient information (or funds) to come your way in order to begin building a digester that will process farmstead waste into methane to heat your house, power an engine, or perform some other fuel-related task. On the other hand, maybe you need to be convinced (or to convince someone else) that waste materials really can be converted into an energy source. This simple, low-cost experiment will help you to become familiar with some of methane's characteristics, and to understand how the fuel is produced.

All a person needs for this demonstration digester are a gallon cider jug, some sort of gas holder (a heavy-duty plastic bag will do), and—from the chemistry lab—some rubber tubing, a couple of tubing clamps, a two-hole rubber stopper, glass tubing and a glass "Y".

The first step in constructing a mini-methane-generator will be to make a manometer. This is a U-shaped tube, partly filled with water, that will let you know when the little digester is producing gas, indicate the pressure of that gas, and act as a safety valve, since excess pressure will blow the water out of the manometer. Any chemistry student should be able to show you the proper way to heat and form glass tubing to the desired shape.

The four-inch manometer dimension shown in the drawing should be considered a maximum for both practical and safety reasons. Filling the tube with water to such a depth will yield eight inches of pressure, which is more than sufficient. Gas appliances usually operate on pressures of less than eight inches, and there's no reason for you to risk blowing your jug apart with gas compressed beyond that amount.

Once the manometer is completed, make a "burner tip" by drawing out a piece of glass tubing in the approved manner (again, with help from that chemistry student if you've never formed glass tubing before). Make the tip quite long to preclude the possibility of a backflash. Then attach the stretched-out burner to one arm of the glass "Y" with a short piece of rubber tubing on which a clamp is placed to act as a valve.

The other branch of the "Y" feeds directly to the gas collector through a longer section of rubber tubing (also fitted with a clamp). A good collector is a collapsible polyethylene milk bag taken from a cafeteria-type dispenser. The cardboard cartons that fit inside such dispensers are thrown out after one use and you'll find that each box contains a bag liner. Fully inflated, the bags are somewhat larger than king-sized pillows. Wash one out, roll it up to expel the air inside, and hook it to your "Y".

Now you're ready to place some manure in the jug. The best type appears to be a mixture of droppings and litter from a chicken barn, but if that's unavailable, try something else like horse manure. The most efficient formula is 30 parts of carbon to one part nitrogen, but don't be concerned about that at first. The first objective is to get the methane experiment moving.

Mix the manure with water to form a slurry and pour it into the jug. (The narrower the container's opening, the more humbling the experience!) Fill the jug to about four inches below the stopper, because there'll be some initial foaming, and you want to keep it out of the tubing.

The most efficient generation of methane takes place at 90° to 100°F, and if the slurry's temperature drops much below 80°, gas produc-

156

tion will be slow or nonexistent. You'll have to provide a sufficiently warm environment for the jug, then, if you want it to make gas. Bear in mind, though, that methane—carelessly handled—can explode, and do take suitable precautions in setting up your apparatus.

Start the generator working with all its valves (clamps) closed, and after a couple of days, the water being "pushed" up the long arm of the manometer will indicate that pressure is beginning to build in the jug. This first production is mostly carbon dioxide, which will not burn. Test the gas by holding an ignited match at the tip of the burner and opening its clamp. The amount of gas in the manometer is sufficient for such a trial, although—as stated—the carbon dioxide will not burn.

Continue the tests until a match held at the burner tip does ignite the escaping gas. This may take a couple of weeks or more depending upon the acidity of the slurry in the jug.

Eventually, incorrect acidity levels will correct themselves, and the model generator will begin to produce methane. When it's evident that such production is underway, open the clamp to the gas collector, and you'll be in business. Methane

production, depending on temperature, should last from one to three months.

And what can you do with the gas? Well, you can burn it off through the burner tip as a graphic demonstration that decomposing organic matter really does produce usable fuel. The quantity this digester yields is too small for much else. To increase the pressure of the escaping gas (and, thereby, the spectacular nature of the resulting flame), place one or more bricks on the collector bag when you try this stunt. The manometer, of course, will faithfully indicate the pressure your gas reaches during such a demonstration.

Once the thrill of watching the flame passes, disconnect the collector bag, take it outside and expel the remaining methane. The residue of slurry in the jug is an excellent fertilizer, or you can use the liquid and some of the solids to seed your next batch of waste and thereby hasten its production of gas.

Although this experiment is imprecise and yields only a small quantity of methane, it will familiarize you with the digestion process and, possibly, encourage you to investigate the construction of larger-scale generators that *will* produce usable quantities of gas.

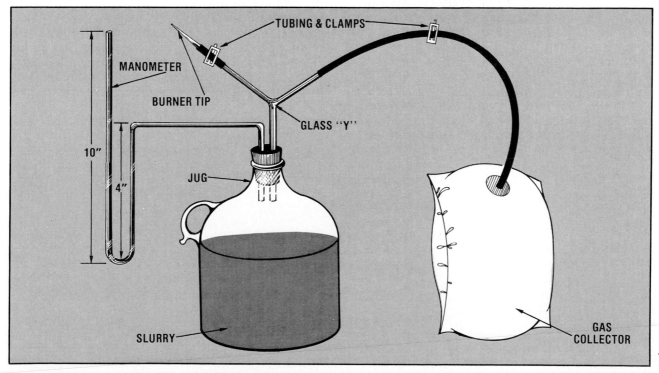

THE METHANE MIRACLE

A Backyard Methane Digester

The three-chambered unit pictured here is capable of supplying between 10 and 18 cubic feet of biogas daily, which is enough to cook an evening meal for an average family.

The large-scale production of methane from animal and vegetable waste offers a possible solution to several societal problems, but what about the individual who may have—each week—only a few stalls of manure and a small amount of vegetation to feed a biogas digester? For such a modest operation, a homemade backyard gas generator can be assembled quickly, easily, and inexpensively.

The heart of this low-cost methane maker is basically a drum within a drum: The outer container serves as a holding tank and digestion chamber, while the inner vessel, which is invert-

ed to capture the gas, acts as a storage canister. To insure that this methane-filled gas-bonnet doesn't tip sideways and release its contents (and to counterbalance some of the upper tank's weight), a cable and counterweight system is hung from a freestanding framework to guide and support the inner barrel. Additionally, because the production of methane gas can take place only within a temperature range of 70° to 104°F, the digester also incorporates a simple but effective passive solar heating system, consisting of a corrugated fiberglass-reinforced plastic (FRP) jacket wrapped around the black-

IT'S GOOD WHEN IT'S ROTTEN

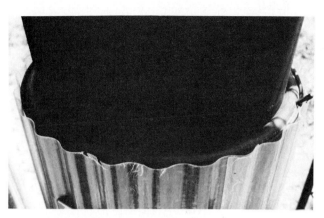

A BACKYARD METHANE DIGESTER

ACETYLENE HOSE

1/8" CABLE

HOIST RING

1/8" X PIPE TO 1/8" HOSE BARB

1/8" X 1/4" PIPE BUSHING

1/4" FEMALE NEEDLE VALVE

1/4" CLOSE NIPPLE

1/4" PIPE COUPLER

20" X 36" WELL TANK

1.75" X 20" BICYCLE INNER TUBE

23" X 35" 55-GALLON DRUM

SEAL HOLES WITH THREADED PLUG

PACK AIR GAPS BETWEEN GLAZING AND DRUM WITH FOAM RUBBER PLUGS FOR COLD WEATHER OPERATION

THREE 26" X 36" CORRUGATED FRP SECTIONS GLUED AND FASTENED TOGETHER

REMOVE BOTTOM FROM INNER TANK

coated outer drum. In the summer, this thermal casing probably isn't needed, but during the colder months, it collects heat and substantially reduces heat loss.

Begin construction of this homebuilt biogas digester by gathering all the materials needed for the project. Tool requirements include a power drill with an assortment of bits, a selection of wrenches, a pair of locking pliers, a pair of shears or a utility knife, wire cutters, solder, a propane torch, a plumb bob, a saber saw with a metal-cutting blade, and an oxyacetylene torch.

You'll also need one or more 55-gallon drums (depending on how many digesters you plan to use in your system) and an equal number of water well pressure tanks. The beauty of such a tank is fourfold: It is easily acquired in most rural areas, is a perfect fit within a standard barrel, is lightweight, and usually is unrusted except at its base, which will be removed anyway. All the other necessary hardware is itemized in the materials list. The type of frame used to support the gas bonnets can vary. It's possible to fabricate one from 1" EMT and 1/4" reinforcing bar, but you can use scrap pipe, 4 X 4's, or even a cast-off swing set with equally good results.

Start by removing the head from the barrel and painting the container's outer surface with high-temperature flat-black enamel. Next, prepare to "wrap" the drum with a flexible glazing (the corrugated FRP used for patio roofs is an ideal choice) by cutting a 26" X 120" panel into three 26" X 36" sections. Then glue and either pop-rivet or bolt the 36" edges of the pieces together with a total of three "one rib" overlaps to form a snug thermal coat that can be slipped over the painted barrel.

THE METHANE MIRACLE

A Backyard Methane Digester (continued)

At this point you'll want to prepare your gas bonnet. First, turn the tank up on its domed lid and remove the bottom with either your saber saw or a cutting torch. Then check the vessel carefully for holes or unwanted openings and seal all possible leaks with welds or threaded plugs. Next, set the tank upright again and fasten a hoist ring to the center of the container's cap. Adjacent to the ring, drill a 3/8″ hole through the lid and weld a 1/4″ pipe coupler over the opening. Finally, give the entire assembly a coat or two of flat-black paint.

You're now ready to set up the miniature bio-gas production facility. Select a level spot that gets plenty of sun and erect the support frame so that one of its "broad" sides faces south. If you are building more than one unit, equally space the 55-gallon digestion chambers beneath the ridgepole of the frame, and then hang a plumb line or a weighted string from that pole so that each barrel can be perfectly centered below the hanging marker. Once you've located the centers, note each spot where the bob is fastened to the pole and attach pulleys at these points.

Next, mount another rope pulley at the end of the overhead structure, thread the appropriate section of cable through the grooved wheel, place the gas bonnet within its container, and fasten one end of the cable to the hoist ring at the top of the tank with a cable clamp. Attach counterweights such as cinder blocks, concrete-filled pipe, rock-laden buckets, or iron bars to the opposite end of the cable, but remember that the weights shouldn't be so heavy that they will lift the steel vessel. They're intended only to keep the lid square and to help when the container has to be raised for any reason.

Cinder blocks and cables through pulleys keep the gas bonnets upright and counterbalanced. Methane gas is drawn off through a hose connected to the tank top with a needle valve and a pipe coupler. Corrugated glazing surrounds each barrel to retain solar heat.

IT'S GOOD WHEN IT'S ROTTEN

The gas handling tubes consist of a combination of welding hose, needle valves, hose barbs, and copper pipe. Using a close nipple, thread a needle valve to the 1/4″ pipe coupler on the gas bonnet. Then fabricate a manifold from copper tubing by drilling a 3/8″ hole through the pipe for each digester and soldering the threaded end of a hose barb to it. Cap one end of the tube, attach a needle valve at the other end by using a sweat-to-pipe fitting, and secure the manifold to the support frame's ridgepole with cable ties.

Finally, connect the required length of acetylene hose from the gas collection chamber to its fitting on the manifold, making certain that the tubing has sufficient slack when the bonnet is at its lowest position, and run whatever length of hose you need to stretch from the manifold to the spot where you are planning to use the home-made biogas.

This small-scale biogas generator is a batch-fed digester. In other words, it must be filled at the beginning and emptied at the end of each production run. During the warmer months, you can expect methane production to start approximately two weeks after you load the first batch of slurry, and good quality gas should continue to be formed for several more weeks. This three-chamber unit is capable of supplying between 10 and 18 cubic feet of biogas per day, which is enough to cook an evening meal for an average family. However, because of the digester's modular design, more chambers can easily be tied into the system in order to increase the production of gas.

Remember, too, that every methane digester has a by-product: nitrogen-rich fertilizer left over from the anaerobic decomposition process. When you consider the cost of building a three-drum setup, you'll probably agree that this may be one of the best ways yet devised to clean out a barn.

You must also keep in mind that biogas is dangerous, especially mixed with air in a closed container, and should be handled with caution. The best practice is to vent off the first part of each batch of methane gas into the atmosphere rather than to attempt ignition of the oxygen-contaminated fuel in any appliance.

LIST OF MATERIALS

20″ X 36″ well tank	(2) 1/8″ pipe to 1/8″ hose barbs
23″ X 35″ 55-gallon drum	1/8″ to 1/4″ pipe bushing
26″ X 120″ corrugated fiberglass reinforced plastic glazing	acetylene hose with 1/8″ bore
(4) 1/8″ bolts or pop rivets	(2) lengths of 1/2″ copper tubing
1/4″ X 2″ hoist ring	1/2″ copper tubing pipe cap
1/8″ cable	1/2″ copper tubing 90° elbow
(2) 1/8″ cable clamps	1/2″ copper tubing coupling
(2) 1/8″ rope pulleys	1/2″ copper tubing to 1/4″ pipe adapter
threaded pipe plugs	(4) cable ties
1/4″ pipe coupler	1.75″ X 20″ bicycle inner tube
1/4″ close nipple	foam rubber scraps
(2) 1/4″ female needle valves	high temperature flat-black paint

THE METHANE MIRACLE

Made in China

With a sizable rural population spread over a vast territory, the need for a localized source of energy has long been felt in The People's Republic of China, where development and the use of modern equipment has been spreading to that Asian land's decentralized, nonurban regions. For nearly 30 years, the Chinese have been experimenting with production of methane (biogas) from household and farm waste.

The research—based on the country's long-standing tradition of composting all available human, animal, vegetable, and crop refuse to make organic fertilizer—has led during the past decade to the development of about seven million small biogas-producing plants located throughout China, with many of the facilities situated in Sichuan Province in the nation's southeastern region.

Understandably, the switch to biogas energy has had an enormous impact on the economic growth of this rapidly "modernizing" country. In the past, rural inhabitants had to either forage for firewood or purchase coal or charcoal to meet their cooking needs. These practices not only stripped the countryside of forests and the fields of organic residue, but also put a strain on mining and transportation facilities and were highly labor-intensive.

From an agricultural standpoint, the production of methane fuel also makes a lot of sense. Obviously, there's only a short-term gain to be realized by burning crop residues for fuel, but the closed fermentation process typically used in biogas "pits" allows the full potential of cull and waste material to be realized.

The official figures for various Sichuan-based methane plants show that the ammonia content of an organic fertilizer mix which has been fermented for 30 days in an enclosed digester is increased by over 19%, while reports from other provinces indicate that gains of up to 160% are possible. This is opposed to outright losses of 82%—by way of evaporation—of the same kind of raw material's potential ammonia when it's left in compost piles. The usable phosphate content of the "biogassed" waste also showed an improvement: 31% over that of conventionally composted mixtures.

Furthermore, when the digested organic fertilizers were applied to existing fields (both the liquid effluent and dried sludge forms were used), crop yields rose noticeably, ranging from an average rice harvest increase of 7.8% to a net wheat production gain of 15.2%.

Perhaps even more significant—especially in view of the lack of information available in our own country on the subject—is the Chinese attitude toward human waste and its role in the ecological cycle. Naturally, owing to the fact that no less than 16 different pathogenic bacteria can be present in human excrement, the disposal of such material must be handled cautiously. The careless dispersal of raw sewage (and even of treated matter that hasn't had adequate time to "detoxify") often results in the spread of infectious disease.

However, if properly processed, human waste can greatly benefit soil fertility, and in some regions of rural China, where sanitary facilities have generally been primitive, at best, a conscientious program of feces disposal in conjunction with biogas manufacture has proved to be the answer to energy and fertilizer shortages and to many community health problems as well.

IT'S GOOD WHEN IT'S ROTTEN

Sichuan scientists have proved repeatedly that a conglomerate of human and animal wastes contains, after a four-week or longer anaerobic digestion period, from 90 to 99% fewer parasite eggs than untreated manure does. A great number of harmful pathogens are destroyed in a matter of hours after the material enters the biogas pits, but the more tenacious parasites (including hookworms, tapeworms, and *Schistosoma*) endure for extended intervals. It's only by means of the relatively high-temperature (85° to 104°F), airless environment of a methane generator that such undesirable organisms are eliminated, allowing the spent "slurry" material to be spread safely on fields used for food production.

Naturally, no one specific biogas plant plan is used throughout China, and even those of similar design must be modified to fit the terrain and soil type of the area in which they're used. Whether the methane plant serves a rural household of several people or a commune of several hundred, it is constructed with a number of important criteria in mind.

First, the plant is invariably located in the ground, which [1] saves valuable farmland, [2] provides thermal insulation both to promote digestion of the manure substrate and to discourage the structural cracking that can result from swings in temperature, [3] allows the substratum soil to serve as a structural aid, and [4] takes advantage of gravity to assure completely trouble-free feeding.

Site selection is also carefully considered. Ideally, a pit is located near a family's living quarters (especially in the case of the household-sized digester) so the latrine can be situated fairly close to the house. But at the same time, the

unit should also be near—or in some cases under—the pigsty, to provide convenient manure loading. (The floors of such stalls typically slope into a trough that leads directly to the digester inlet, permitting the waste to be washed downward daily.)

Other site considerations include the need to place the fermenting pit away from an area with

A digester for a typical family is compact and accessible, allowing for easy periodic cleanouts.

163

THE METHANE MIRACLE

Made in China (continued)

trees and large roots, *and* to position it in such a way that the soil above the chamber will receive sunlight, allowing the unit to take advantage of direct solar gain. The composition of the foundation material is also critical, since the presence of expansion clay, mixed soils, groundwater, or solid rock could call for the use of a different construction method or relocation of the pit.

Finally, each unit must be sized to handle adequately both the incoming waste and the fuel needs of the family or community that it serves. The Chinese have found that a household with three to seven persons, and perhaps a pig or two, requires a digester of 200- to 425-cubic-foot internal volume, based upon the assumption that each family member uses about ten cubic feet of gas—for cooking and lighting—per day and that the daily yield of fuel from a typical small-scale pit digester is approximately 15% of the volume of the fermenting liquid.

In order to cut the cost of constructing the units while conserving materials, the builders try to use what's available locally when fabricating each pit. Systems are usually built by hand, and

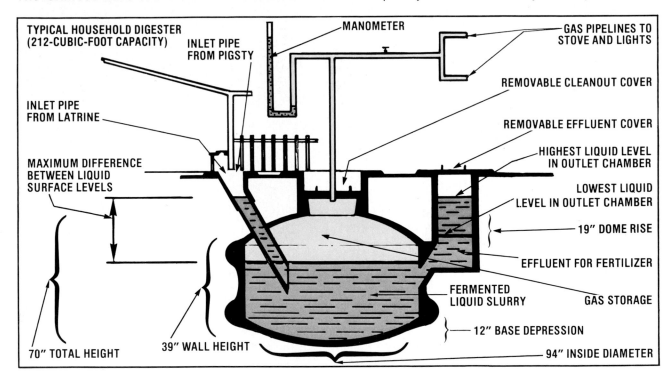

TYPICAL HOUSEHOLD DIGESTER (212-CUBIC-FOOT CAPACITY)
INLET PIPE FROM PIGSTY
MANOMETER
GAS PIPELINES TO STOVE AND LIGHTS
INLET PIPE FROM LATRINE
REMOVABLE CLEANOUT COVER
REMOVABLE EFFLUENT COVER
MAXIMUM DIFFERENCE BETWEEN LIQUID SURFACE LEVELS
HIGHEST LIQUID LEVEL IN OUTLET CHAMBER
LOWEST LIQUID LEVEL IN OUTLET CHAMBER
19" DOME RISE
EFFLUENT FOR FERTILIZER
FERMENTED LIQUID SLURRY
GAS STORAGE
12" BASE DEPRESSION
70" TOTAL HEIGHT
39" WALL HEIGHT
94" INSIDE DIAMETER

IT'S GOOD WHEN IT'S ROTTEN

This excavation will serve as a biogas
pit. Large methane plants are
sometimes covered with cement ponds
where aquatic vegetation is cultivated.

THE METHANE MIRACLE

Made in China (continued)

indigenous rock, lime, clay, and homemade brick are incorporated—whenever possible—to provide structural support or to cement components together. (It's interesting to note that, over the years, Chinese workers have developed a variety of concrete grades made from coal and lime slag, crushed brick and tile, ash, clay, cinder, or sand—in short, they use many substances which we would often consider waste—and feel that the resulting construction material compares favorably with the "real thing".)

Normally, a pit is first excavated, and then the digester's masonry "shell" is laid up around a wooden or steel mold. Occasionally, though, the container is cast right in place. In any event, a typical unit consists of a central liquid-manure and gas storage tank connected to a slanted inlet pipe that extends into the slurry. Opposite this—and separated by a small retaining wall—is an outlet vat containing about one-tenth the volume of the main tank. Effluent fertilizer can be withdrawn from the vat through a covered hatch. The gas itself travels along a pipe fastened to a removable cover at the top of the domed tank lid and is used as needed (a simple water-in-tube manometer is employed to measure internal pressure). At some locations, ponds are built on the area directly over the tank to seal potential gas leaks, store solar heat, and grow aquatic vegetation.

The Chinese digester's simple design maximizes its utility. Because the vat is broader than it is high, it is not likely to interfere with deep groundwater supplies, is easy to excavate a site for, and has more potential for gas production than would a taller unit because of its greater surface area. Moreover, the steeply angled inlet

pipe can't clog, and the force of the fresh waste material entering the system stirs up the fermenting matter, helping to speed decomposition. Additionally, the position of the outlet chamber allows only "sanitary" effluent to escape, since parasitic eggs and bacteria settle to the bottom of the tank where they're eventually destroyed. Finally, the cleanout cover allows for the necessary semiannual inspection and manual emptying of the entire vat.

Depending on the size of the digester, the gas produced can be used in a number of ways including burning it in lamps for light and in stoves for the preparation of meals. But the product of communal methane plants is often used to feed gasoline or diesel engines which have been inexpensively converted to burn methane, either in conjunction with, or as a replacement for, conventional liquid fuels. (In the case of gasoline powerplants, a feed pipe equipped with an adjustable valve is tapped into the carburetor body. On diesel engines, a similar arrangement supplies the intake manifold with biogas, which is ignited in the cylinder by a small amount of diesel oil.) Such powerplants usually drive generators which furnish electricity for lights and machinery.

If the example set by the Chinese can be duplicated in other nations—including, perhaps, our own—there seems to be little reason why at least some of the world's poor can't enjoy better soil, partial energy self-sufficiency, and freedom from disease . . . at almost no cost!

IT'S GOOD WHEN IT'S ROTTEN

A Serious Last Word

The most important consideration when dealing with methane is the safe storage of this flammable material. And the most important safety precaution in a system for storing the gas is never to mix methane and air and then ignite it! A mixture of gas to air between 1 in 14 and 1 in 4 (between 7% and 25% gas) is explosive, and no chance should be taken, even if your mix is just outside, but still close to, these limits.

The second vital safety measure is to pipe and store the gas under pressure. It can be very light pressure, such as that provided by a floating gas holder, but there must always be enough positive pressure to prevent air from entering the digester, piping, and storage tanks.

The "wrong" drawing shown here illustrates a dangerous way of storing methane in which both these rules are ignored. There is no way to know the ratio of gas to air in such a tank. (The gas may be lighter or heavier than air, depending upon the stages of decomposition in the digester.) With such a setup, the gas being introduced at the bottom of the tank will not rise and force the air out as would water filling a container. Instead, the methane will mix *with* the air (and the oxygen in it) to create an explosive combination.

If the mixture is then ignited at a burner outside the tank, it would explode back into the tank. And if the gas/air mix were used in an engine, the engine could backfire and blow up the tank. There is simply no way that this design can be used safely. Furthermore, apart from the extreme danger present, there is only one way to withdraw the gas mixture from a fixed capacity tank, and that is to fill the container with water, thus forcing the fuel out. And this method is just not practical in most cases.

A final example of the dangers of methane gas is to be found in underground mines. Odorless methane seeps into the shafts to mix with the air confined within this enclosed space. In such a situation, a flame—or even a spark—results in a disastrous accident and the loss of miners' lives.

There are, however, correct ways to store methane. A well-designed floating gas holder, guided so that it can rise and fall according to the amount of methane entering or leaving the tank, will stay free of oxygen. It will be so safe that if it should be ruptured by lightning, an accident, or even by an incendiary device, the gas would merely burn. Since the methane would have entered the container oxygen-free, and no air was permitted to enter the gas holder, digester, or piping at any time, the gas would not explode.

Another safe way to store methane is in pressure cylinders. These, by law, must be authorized, tested pressure vessels and must be certified by an approved authority who will present a certificate of the tests conducted. The examination must be repeated at regular intervals.

Stored improperly, biogas can be deadly. So care for it sensibly, follow the rules, and enjoy your inexpensive and efficient source of energy safely.

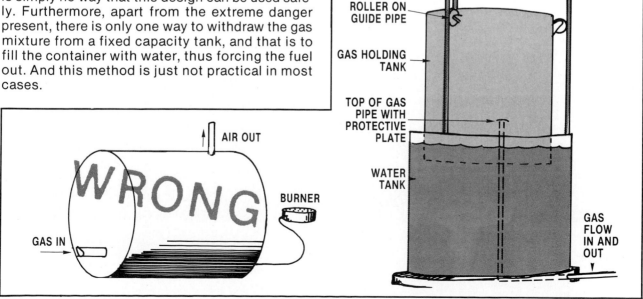

167

FUEL FROM THE FOREST

Woodstoves have brought low-cost warmth to homes in recent years, but with the revival of the popularity of these tree-fueled appliances have come several reasons for concern. Safety is the top priority for most wise stove owners, and standards for installing the heaters in ways that would minimize the chance of a house fire have been devised.

To reduce the amount of heat that's lost up the chimney, stove designs were sought that burned the fuel more completely. And, happily, such developments brought with them a reduction in the amount of pollution leaving the chimneys and entering the atmosphere.

In addition to the growing use of newly designed, modern woodburners, demand has grown for the antique stoves that were used to heat or cook in most homes before the convenience of natural gas or electricity was commonplace throughout the country. A great many of these have been restored, and others are undoubtedly waiting in barns or sheds for new friends and new homes.

And, while all woodburners use a fuel of the same name, all wood is hardly the same. Knowing the qualities of one's fuel, such as the amount of heat it will produce, its cost, and its availability, is important to anyone considering switching to this fuel from the forest.

FUEL FROM THE FOREST

Choose the Right Tree

With the recent astronomical rises in the costs of gas and oil as heating fuels, many people have begun depending on fireplaces and woodstoves to provide at least a portion of the heat for their homes, and to limit their dependence on high-priced, nonrenewable energy. And along with this rise in popularity has risen concern over just what kinds of wood make the best fuel in home-heating applications.

Most woodstove and fireplace owners have probably spent at least some time talking about the different types of wood they consider best for heating or have at least heard someone extol the virtues of one variety over another, usually recommending the more dense woods, which do offer significantly more heating value—and therefore burning time—than do less "heavy" varieties. The reason for this is simple: A given volume of a dense wood (hickory, for example) weighs considerably more than does the same volume of an airy species such as pine. Since a pound of dry wood has approximately 8,500 Btu of heating capacity, a cord of heavy wood contains more energy content than does a cord of lighter wood.

Consequently, whether you're cutting your own wood or buying by the cord, it's important that you know what you're getting. Although the age of a tree, and even the region of the country in which it grows, can affect a specie's appearance, the accompanying comparative chart (see page 172) and the identification photos should help in selecting the best trees for firewood from the types which grow in any given area.

When using the chart to help you make this important decision, be sure to weigh the advantages and disadvantages of each variety. For instance, you can see that hickory has twice the heat value of pine and therefore should be worth twice as much to you. However, the popularity of hickory and oak as stove fuels often drives up their prices, so an extensively-seasoned (for at least one year) cord of black locust—which has roughly the equivalent heating value, but lacks the reputation of the more notable burnables—could prove to be a better investment.

On the other hand, though you can often haul away a load of sycamore for a nominal charge, splitting a cantankerous chunk of *that* stringy stuff could be enough to drive many folks back to oil furnaces (and that's unusual at today's fuel prices!) Of course, there's some argument on the subject of splitting wood: *Most* maul swingers find sycamore—which is often used for chopping blocks—difficult to crack, while others claim that it's just a matter of having the right wedge angle.

But there are still other factors that might well influence a person's choice of firewood, especially if that individual is cutting and hauling the timber. Among these are the distance to the woodlot or forest, cutting fees, and the accessibility of the wood by road. A good "free" stand of oak or hickory might become very "labor expensive" if the logs have to be carried any great distance by hand or need to be hauled a long way by vehicle.

The best time to bring in a winters' supply of firewood is also open to discussion, but it's without a doubt easiest to identify and mark the suitable trees when they're in full leaf. Then, once the future firewood is tagged, cutting can wait until a cooler time of year . . . for the comfort of both you *and* your chain saw.

ASH

ASPEN

AMERICAN BEECH

GRAY BIRCH

BLACK GUM

PIGNUT HICKORY

CEDAR (JUNIPER)

BLACK LOCUST

CHESTNUT OAK

RED OAK

WHITE OAK

WHITE PINE

SASSAFRAS (MALE)

SYCAMORE

TULIP TREE

BLACK WALNUT

FUEL FROM THE FOREST

Choose the Right Tree (continued)

FIREWOOD FACTS		
WOOD	**SPLITTING**	**HEAT VALUE[1] (BTU) PER CORD[2]**
ASH, Green	fair	24 million
ASPEN, Bigtooth	easy	17 million
BEECH, American	hard	27 million
BIRCH, Black	easy	27 million
BIRCH, Gray	easy	26 million
BLACK GUM (Tupelo)	hard	21 million
HICKORY, Pignut	easy	32 million
JUNIPER (Eastern Cedar)	easy	20 million
LOCUST, Black	easy	30 million
MAPLE, Red	easy to fair	23 million
OAK, Chestnut	fair	28 million
OAK, Red	fair	27 million
OAK, White	fair	29 million
PINE, White	easy	15 million
SASSAFRAS	easy	20 million
SYCAMORE	hard	21 million
TULIP TREE (Yellow Poplar)	easy	18 million
WALNUT, Black	easy	24 million

[1] Subtract efficiency losses for stove
[2] 80 cubic feet

THE FIREPLACE

The Right Way to Build a Fire

While the standard fireplace is hardly *efficient* when it comes to providing heat, its output can be dramatically increased—possibly even doubled—by merely altering the way in which the fuel is stacked on the grate.

The conventional method of piling logs lengthwise in the center of the grate forces the fire to burn between the back wall of the fireplace and the wood, and the fuel itself blocks much of the radiant heat produced by the blaze and prevents it from reaching the room it is meant to heat.

A more effectively laid fire can correct this problem by putting the flames in front of the logs (see page 174 for photos). Start by placing some kindling in the center of the grate. Then lay one split-out section of log across each end of the stack of kindling and perpendicular to the back wall of the fireplace. The logs should be slightly farther apart at the front than they are in the back. They should actually touch the back wall of the fireplace but should not extend more than an inch or so beyond the front of the grate (if they stick out too far, they can "throw" smoke into the room as the fire burns).

Next, lay one or two pieces of wood across the two side logs and push these last chunks all the way back so the rearmost piece of wood is shoved right up against the rear wall of the fireplace. The second log (if there is one) is crowded against the first.

This system creates a "firebox" that's entirely open across its front and thus allows a great deal of energy to radiate directly into the room as the fuel burns. With this kind of setup, the flame draws fresh air up through the bottom of the grate quite freely, because that grate isn't covered by tightly packed logs.

Wood stacked in this manner is easy to light too, but first open the damper and check for a draft flowing up the chimney. If the flue isn't drawing air, twist a sheet or two of newspaper into a tight rope, light it, and hold it up into the vent. This flame will warm the air and cause it to start rising up the chimney. That's the signal to light the main fire in the usual way with twisted papers or a "fuzzy stick". (*Never* use gasoline or any other flammable liquid!) In just a few minutes, you'll be enjoying all the heat that your fireplace normally puts out ... *plus* a good deal more that ordinarily "goes up in smoke".

Eventually the one or two top logs will burn through in the middle and fall down into your wooden "firebox". Just replace them with fresh pieces, and if you want a hotter blaze, shove some smaller chunks right into the fire, positioning them parallel to the side logs. Those side pieces don't burn through very readily: Instead, their back ends just seem to scorch away. When that happens, they can be replaced with new splits of wood, and the old, partially-burned sections can become new top logs for the fire.

In addition to increasing a fireplace's efficiency, this method of laying a fire has another distinct advantage: It allows the burning of greener and moister wood (which can be "cured" as part of the firebox before becoming part of the blaze) than would normally be feasible.

Whether a home's fireplace serves as a major source of heat or is merely used to take the chill from exceptionally frigid winter nights, careful stacking of the logs can glean nearly twice the hearthside warmth your fires have produced in the past, while also stretching the life of your woodpile.

FUEL FROM THE FOREST

The Right Way to Build a Fire (continued)

Lay the kindling and logs to form a firebox within the fireplace, and you'll gain additional heat while making your woodpile last longer.

THE FIREPLACE

A 90% Efficient Fireplace?

The 90% efficiency of this compact fireplace is due to its cast-masonry body, which stores and liberates heat better than does its metal counterpart. The convoluted stovepipe furnishes a quick blast of heat to the room after the initial lighting.

The "rebirth" of wood heat has been a real education for a good many people who grew up thinking that indoor warmth is created by a turn of the thermostat. Not only has the growing popularity of fireplaces and woodstoves prompted a great many folks to put a time-and-sweat, rather than a dollars-and-cents, price tag on their fuel, but it has also stirred up curiosity about the sort of performance that can be expected from the substitute furnaces.

Surprisingly enough, the folks who have looked into such matters have found that the efficiency of the woodburners (calculated by measuring the amount of heat vented into the atmosphere, subtracting *that* figure from the total number of Btu available in the given weight and kind of wood used, and converting the result to a percentage) can range anywhere from below 0% (with a poorly designed fireplace) to a high of perhaps 80% (in a good airtight stove).

In an effort to learn why some heaters work better than others, and with the eventual goal of building an effective, affordable woodburning device, two New Mexican researchers began testing various stove designs.

The designers learned that the New Mexico Energy Institute, a research organization dedicated to the promotion of alternative energy at the University of New Mexico at Albuquerque, was conducting a workshop project on "Russian fireplaces", and these devices were testing out at 90% efficiency! The excellent performance of these unusual heaters stems from two characteristics: They contain enough masonry to provide a massive heat sink, and the interior flue path of each unit is arranged in a serpentine pattern so that hot waste gas has plenty of time to transfer

FUEL FROM THE FOREST

A 90% Efficient Fireplace? (continued)

its thermal energy into the walls of the stove. Unfortunately, the typical Russian fireplace weighs about 11 tons, stands seven feet tall, and can cost a couple of thousand dollars.

However, using information gleaned from the university's seminars, and with the aim of remedying the disadvantages of the design, the two designers created a five-piece, cast-clay firebox that weighed only 280 pounds, cost less than a conventional woodstove, and took up an easy-to-live-with 18″ X 18″ X 27″ space. Performance tests done on the compact creation showed it to be virtually equal in efficiency to the massive Russian fireplaces.

The secret of this small stove's success is not an intricate interior labyrinth, but the fact that cast masonry stores and liberates heat better than does metal. The New Mexican stove was built of a refractory clay that's [1] easy to work with, [2] strong, [3] able to withstand 2500°F temperatures, and [4] doesn't require kiln firing.

To build the "oven", first make a mold using 3/4″ plywood and 2 X 4 braces, and cut it in half horizontally. Next, line the form with thin-gauge sheet metal and cover the inner "plug" in the same manner. Since the rear section of the firebox supports the flue pipe, you must devise a reinforced cardboard bung to provide a hole for that piece of exhaust hardware when casting the "aft" portion of the stove. Then work up a mold for the simple arched slab that forms the back wall of the firebox.

The trickiest part of the stove's construction, its designers found, was the actual casting and curing. The refractory clay must be mixed with enough water to give it the consistency of wet sand that will hold its shape when squeezed. Accordingly, it's best to be thorough when dampening the mix, using more water than the manufacturer recommends to make the strongest blend.

Since the "mud" is too stiff to pour, it has to be carefully spooned into the molds (which ought to be coated with a layer of grease beforehand to make them easier to remove later) and firmly tamped—with a piece of 1 X 2—after each addition, to be sure all of the corners are filled. The mixture starts to set in about 20 minutes.

When all the molds are full, excess clay can be skimmed off the tops and a sheet of plastic placed over them. The castings should be allowed to stand this way for one week, at a temperature

The pattern for the wooden mold is transferred from a cardboard template to plywood.

THE FIREPLACE

that's higher than 50°F. Then—since proper curing requires that the clay be kept slightly damp for at least 14 days after forming—*gently* remove the molds after the first week and seal each "green" piece into a plastic garbage bag. Allow the sections to finish their initial curing in that moist environment. Finally, remove the waterproof coverings and sun-dry the parts for another two to four weeks, or until they're completely air cured (naturally, exposure to rain and condensation should be strictly avoided during this final stage). While you're waiting for the molded components to harden, it would save time to order or fabricate

an adjustable, vented steel door to fit the front of the firebox.

The New Mexican stove builders have assembled their creations both over beds of bricks and on poured concrete slabs. To hold the various components together, they used thin layers of refractory clay as mortar in the joints, though they say refractory cement will work, too. (Another two weeks of air-drying is needed to cure the seals.)

The flue installation is very important to the stove's ability to transmit a maximum amount of heat properly. In a standard room with seven-to eight-foot ceilings, a 16-foot length of 6" pipe—

Once the outer mold is sawed in half horizontally, the two forms are ready for covering.

FUEL FROM THE FOREST

A 90% Efficient Fireplace? (continued)

equipped with an airtight damper and arranged in a serpentine pattern—seems to work best. With too much stovepipe, there may not be enough hot exhaust flow for the stove to draw properly. Also, the firebox and flue stack should be 36 inches away from all combustible surfaces, and a ventilated wall thimble should be used.

There's a means of keeping an eye on the creosote that's effective, inexpensive, and simple to install: An inspection port—about 3″ X 3″—is cut into the stovepipe, covered with a short section of the same diameter stovepipe held in place by a small spring, or a bolt and wing nut.

When the assembly is completed and checked, it's ready for its initial fire. In order to complete the curing and conditioning process of the masonry, fuel for the first five burnings should progress from a single sheet of newspaper to a healthy handful of kindling. Afterward the stove can be used normally.

The heater will, when loaded with a mere five pounds of wood and lighted, give an instant dose of warmth to the room through its stovepipe. An hour and a quarter later, the outer surface of the firebox will reach nearly 200°F. And six hours after that, its masonry "skin" will still be hotter than

A thin-gauge sheet-metal skin is wrapped around and tacked to the inner mold, and then the two halves of the outer mold receive a similar application.

THE FIREPLACE

100°F, though the flue pipe will have long since cooled. The fire should be hot and fast-burning until the coals stage begins, usually after about 35 minutes. Then the flue damper must be closed to allow only a *slight* flow. This mode of operation allows the fireplace to absorb the heat rather than letting it escape up the chimney.

When the New Mexican designers tested their stove, they came up with efficiency figures as high as 92.6%. And, even if this incredible percentage doesn't pan out for everyone, the "ceramic stove" could be about the closest thing to home-heating perfection available.

The metal surfaces of the completed castings frame are coated with oil or grease before packing in the refractory clay. After the castings cure for a week, the forms are removed and the finished body (the photo shows one half) is first damp-cured and then sun-dried.

FUEL FROM THE FOREST

Basic Woodstove Safety

At one time, the safe operation of a wood-stove may actually have been a matter of using plain common sense. But before assuming that the knowledge of campfires and fireplaces will carry through to the installation and use of one of today's complex airtight heaters, be aware that not even the "experts" can agree on the exact precautionary steps prudent woodburning should involve.

Much of the controversy centers on the fact that a compromise between absolute safety and the highest possible efficiency is often necessary, but no one can determine what margin of safety (at the expense of what degree of efficiency) a person should choose.

Far too few consumers know just how important it is to choose a stove of the right capacity for the area to be heated. After all, wood heaters are anything but inexpensive, and financial pressures often force many folks to opt for a small unit, which they often wind up overfiring.

And, since the recognized safety standards for woodstove installation are based on normal operating temperatures, it's possible—though the standards include comfortable margins for error—for an overheated firebox to ignite walls or other combustible materials that are beyond the "safe" perimeter. For example, sustained temperatures of slightly over 200°F can actually cause wood to burst into flame spontaneously.

On the other hand, some shoppers buy top-of-the-line models, assuming that a large stove will provide them with reserve heat. But the dangers of an underfired heater are also serious. Since most owners of large stoves often reduce the output of their heaters by restricting the air supply with a damper—a technique that reduces burn efficiency, makes more smoke, and forms creosote rapidly—buying a stove that's larger than it needs to be can be the indirect cause of chimney fires. In fact, such accidents have been known to happen within a week of the installation of a new stove and chimney.

The best defense against a raging chimney blaze is to close up the stove, shutting off the smokestack's oxygen supply, so it's imperative that any stove be sound and well sealed.

However, if the stove has any leaks around its door or damper, or through the body of the stove itself, attempting to starve a chimney fire can create a disastrous situation.

If there's even a small flow of air to the smoldering creosote, the oxygen will accumulate until the chimney's coating flashes into a blaze. This ignition will quickly consume all the available air, and the fire will expire until the oxygen accumulates again. Repeated flashing—which can happen as often as several times per second—has been known to break apart stovepipes and even masonry chimneys!

All new stoves approved by the Underwriters Laboratories (UL) meet standards of soundness and engineering, but secondhand units often suffer from deterioration, so check any used heater carefully. Also, be certain that the stove is designed to burn wood: Few coal stoves can accept logs, and no woodstove can safely burn coal without an appropriate grate.

The first concern when installing a woodstove is to maintain safe clearances. Follow the specifications in the accompanying tables, and—if space for the stove is limited—use suitable thermal barriers. The National Fire Protection Association (NFPA) specifies using asbestos mill-

THE WOODSTOVE

board—not cement board—and/or steel plate, but because of the harmful effects of asbestos, conscientious stove installers will opt for steel and an adequate air space.

When a woodstove's body is set 18 inches or more above a flammable floor (or even if the framework supporting this surface can burn), line the floor under and near the heater with 24-gauge sheet metal. If the clearance is less, use four inches of hollow masonry *underneath* the metal.

The stove's position will determine the amount of stovepipe necessary to connect it to the chimney. The NFPA specifies that as little stovepipe as possible should be used, but it's generally accepted that every four feet of pipe (up to about 12 feet) that's exposed to inside air space will yield about 10% "bonus" heat from a stove. Uninsulated stovepipe does an excellent job of radiating the heat from flue gases.

However, beyond the obvious danger of someone's being burned by the hot metal surface, there are two other basic hazards connected with using an extended stovepipe. First, as the pipe's length increases, so does the possibility that its joints will break during a chimney fire.

Second is the question of whether the heater's draft is adversely affected by a long stovepipe. The heat of the flue gases is dissipated through the metal, and this reduces the thermal draft effect. But longer sections also tend to *improve* draft by increasing the chimney's volume. The trade-off (as well as the point where one effect overwhelms the other) is uncertain.

The uninsulated stovepipe should clear all flammable objects by the same distances suggested for stoves in the tables of clearances.

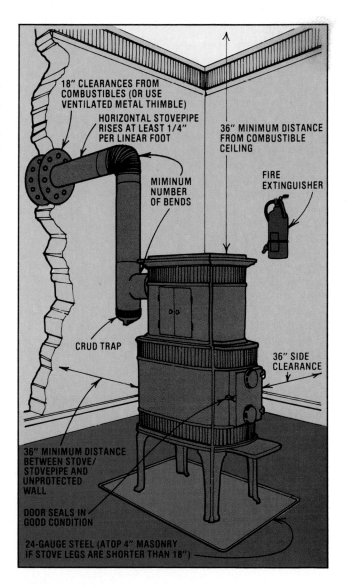

18" CLEARANCES FROM COMBUSTIBLES (OR USE VENTILATED METAL THIMBLE)

HORIZONTAL STOVEPIPE RISES AT LEAST 1/4" PER LINEAR FOOT

36" MINIMUM DISTANCE FROM COMBUSTIBLE CEILING

MIMINUM NUMBER OF BENDS

FIRE EXTINGUISHER

CRUD TRAP

36" SIDE CLEARANCE

36" MINIMUM DISTANCE BETWEEN STOVE/STOVEPIPE AND UNPROTECTED WALL

DOOR SEALS IN GOOD CONDITION

24-GAUGE STEEL (ATOP 4" MASONRY IF STOVE LEGS ARE SHORTER THAN 18")

FUEL FROM THE FOREST

Basic Woodstove Safety (continued)

Again, the necessary space can be reduced by using a thermal barrier. It's also possible to use costly factory-built insulated chimney with clearances specified by the manufacturer, but a great deal of the heat gain from the stovepipe would be lost.

To prevent creosote from leaking from the stovepipe's seams, and also to keep smoke from escaping, the female ends of the pipe should be mounted upward and then each junction should be sealed (except those which facilitate removal of the pipe for cleaning) with furnace cement and secured with three sheet metal screws spaced around the circumference of the joint.

The NFPA also specifies that all horizontally run stovepipe should rise (away from the stove) at least 1/4 inch per linear foot, so that the flue's draft will be aided by the slight upward flow.

When using an extensive run of stovepipe, support the tubing (usually done by suspending it with wires) at least once every six feet. In addition, always try to employ a minimum number of bends and use the most gradual curves possible to reduce turbulence in the exhaust gases.

Remember, too, that even the most skillfully fitted and maintained stovepipe has a maximum life span of three years. After the first season, the pipe should be checked regularly for soundness. A fairly good test can be made by squeezing the pipe. If you are able to crush the walls in your grip, the corrosive creosote has eaten away too much metal for the pipe to be safe.

One useful feature is a crud trap. Where the pipe leaves the stove—in most cases, horizontally—simply add a tee fitting instead of an elbow. Then plug the lower end of the tee with a removable cap fastened with sheet metal screws.

When it's time to clean or inspect the stovepipe, this opening will give you easy access to what is usually the longest straight section.

There are four basic ways to connect stovepipe to a chimney: [a] direct connection to a masonry chimney, [b] passage through a flammable partition into a masonry chimney, [c] entry into a fireplace and chimney, or [d] outlet through a wall or ceiling via factory-built metal chimney.

Any stovepipe inserted directly into a masonry chimney should penetrate to, but no farther than, the inside edge of the masonry liner and must be sealed with furnace cement unless a masonry thimble is set into the chimney to accept a slip-in section of pipe.

Any connector which passes within 18 inches of a flammable partition falls into the second category. NFPA regulation number 211 specifies that—in such cases—the installation include either a vented metal thimble 12 inches larger in diameter than the pipe, or a masonry thimble ringed by at least eight inches of fireproof brickwork. The only other prudent option would be to leave a minimum of 18 inches around the outside of the stovepipe and then close the hole with a flame-resistant barrier.

When using a fireplace chimney for woodstove exhaust, the entry to the fireplace must be sealed, or if the connector joins the flue above the fireplace, the chimney must be plugged below the point of junction. This precaution not only prevents burning embers from falling down into your fireplace and possibly onto your floor, but it also maintains the proper draft for your stove. Ideally, all such chimney entrances for woodstoves should be above the fireplace, but practical considerations often demand that the

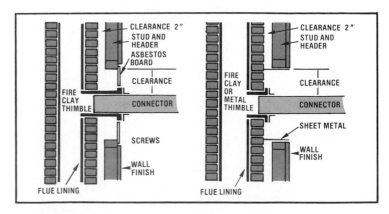

THE WOODSTOVE

connection be made through the sealing partition and into the fireplace itself. To prevent heavy creosoting (especially in any fireplace with a cool external chimney), you should extend the stovepipe so it turns up and into the chimney.

NFPA regulations prohibit the passage of unprotected stovepipe through any floor, ceiling, or fire wall. You may, however, pass pipe through either a wall or a floor/ceiling if a factory-built insulated chimney is used. Without a suitable masonry chimney, this rather expensive piping is the only choice. While costly, the ready-made chimney material is efficient and will improve the draft and reduce creosote accumulation.

When using a masonry chimney, there are several precautions to take before firing up a new stove. For one thing, the cross-sectional area of any chimney should be approximately 25% larger than that area of the stovepipe that feeds it. (Add about 4% for each 1,000 feet in altitude.) A chimney that was built to serve an oil, gas, or coal furnace will probably be much too large for a woodstove, but you can run the stovepipe all the way up the chimney, or (in some cases) place a cap of the appropriate size atop the chimney.

An old chimney should also be checked for soundness before a woodstove is hooked up to it. A technique called "puffing", which involves plugging the top of the flue and starting a small but smoky fire below, is a useful test. The smoke will emerge from any leaks in the chimney liner.

Clean the chimney before firing up your wood-burner if there's *any* significant creosote accumulation. A layer of more than 1/4 inch of the black goo constitutes both a fire hazard and a hindrance to efficient drafting. Either call a chimney sweep or learn to do the job yourself.

What's more, an external (running on an outside wall) chimney may not be worth using. Such structures tend to be less efficient than chimneys that are inside the house because cold air prevents the masonry from conveying much heat into the building. Furthermore, the same cooling effect encourages rapid creosote formation. Unless an outside chimney is in excellent condition and perfectly suited to the stove, it may pay to use a factory-built insulated chimney instead.

Both woodstove safety and efficiency profit from a tall chimney that protects the house from sparks and insures an adequate draft. The outlet of the flue should be at least three feet above the spot where the chimney goes through the roof, and no less than two feet above any portion of the roof that lies within ten feet of it.

Ironically, the actual firing of your stove is the biggest trade-off in the safety/efficiency/convenience game. If it were possible to keep a small fire burning with an adequate supply of oxygen (an open draft of 50% or more), creosote might never be a worry. But the owner would either be up at least three times a night to add wood or wake to a very cold house in the morning. Assuming that you do intend to keep your stove's fire going all night while you remain snug and warm under the blankets, there are a few precautions and specific techniques that help control creosote buildup.

Contrary to popular opinion, burning hardwood (or even an exclusive collection of seasoned wood) will not prevent creosoting. But while hardwoods contain just as much of the creosote-producing substances as do softwoods, green wood can creosote especially heavily if it's not burned correctly. Because the

TYPE OF HEATER	ABOVE TOP	FROM FRONT	FROM BACK	FROM SIDES	
MINIMUM CLEARANCE DISTANCES, IN INCHES, OF FREESTANDING STOVES FROM UNPROTECTED COMBUSTIBLE SURFACES (Do not install in closets or alcoves.)					
				Firing Side	Opposite Side
Room heater	36	36	36	36	
Cookstove w/clay-lined firepot	30		24	24	18
Cookstove w/unlined firepot	30		36	36	18

Minimum Clearance Distance Charts reprinted, with permission, from *Using Coal and Wood Stoves Safely!* Copyright 1978 NFPA, Boston, Massachusetts.

Basic Woodstove Safety (continued)

wood's moisture must be converted to vapor before the fuel can produce the gases which burn, unseasoned or damp wood can lead to inefficient, soot-causing combustion if the timber doesn't have both sufficient draft and an established bed of burning coals to help it get going.

One good technique for producing a relatively clean, long burn is to load your stove with a mixture of partially seasoned and well-dried wood about half an hour before bedtime. Leave the damper open to give the fire a good start until you're ready to retire. Then damp the strongly burning blaze down for the evening.

In most cases, it's best not to buy a stove with a secondary air inlet to help control creosoting. The NFPA feels that such an option usually makes combustion less efficient by introducing cold air into the rising gases. Unless the stove in question has an excellently designed secondary draw that keeps temperatures up in the 1100°F range, there's no way the additional oxygen will help the rising fumes ignite. Cool exhaust gases mean heavy creosoting.

Furthermore, never burn trash in a woodstove. Paper wastes tend to make a very hot fire for a short period of time, encouraging the ignition of any creosote deposits in your stovepipe. And synthetic wastes such as plastic wrappers produce hydrofluoric and hydrochloric acid when they burn. The effect of such corrosives on stovepipe (and the body of the stove) could possibly lead to disaster.

A one-inch-thick layer of ashes at the bottom of your stove acts as a reflector and an insulator to help the heater sustain a hot fire. But when disposing of the wastes, keep in mind that ashes combined with water form a caustic lye solution.

A stove that is not damped excessively and that has a well-designed chimney of factory-built, insulated pipe might go an entire season without needing a "sweep". On the other hand, even a comparatively well-installed system could —when used to hold a fire all night—need cleaning as often as every two weeks.

Thorough maintenance is the only way to make up for using a dirty burn technique or a less-than-perfect flue. After installing a new stove, be sure to check the stovepipe every two weeks for creosoting until you become accustomed to the heater's behavior. (Any deposits over 1/4 inch thick indicate that the pipe needs attention.) It's possible to monitor the accumulation in a stovepipe with some accuracy by tapping on the sections with a metal object. Once a person is used to the ringing sound that a clean pipe makes, the dull thud of a dirty one will be distinctive.

In the long run, sweeping one's own chimney saves money, but it should not be attempted without learning how to perform the job properly and with the right tools. There's no substitute for a chimney brush. Various combinations of chain in burlap bags and caustic chemicals are not particularly effective and can easily damage flue liners. Calling out a sweep is costly, and while such an expense would be tolerable once or twice a year, it will be prohibitive if the system needs a cleaning once a month.

Almost every woodburning household eventually experiences a chimney fire, although the blazes are usually quite mild. There will be no doubt when one occurs, however. The stovepipe will become very hot (perhaps even glowing), and there will be a rush of air through the draft in the

MINIMUM CLEARANCE DISTANCES, IN INCHES, OF FREESTANDING STOVES FROM PROTECTED COMBUSTIBLE SURFACES (Thicknesses shown are minimums.)			
TYPE OF PROTECTION	CHIMNEY CONNECTOR	ABOVE	SIDES AND REAR
[a] 1/4" asbestos millboard spaced out 1"	12	30	18
[b] 0.013" (28-gauge) sheet metal on 1/4" asbestos millboard	12	24	18
[c] 0.013" (28-gauge) sheet metal spaced out	9	18	12
[d] 0.013" (28-gauge) sheet metal on asbestos millboard spaced out 1"	9	18	12
[e] 1/4" asbestos millboard on 1" mineral fiber batts reinforced with wire mesh or equivalent	6	18	12
[f] 0.027" (22-gauge) sheet metal on 1" mineral fiber batts reinforced with wire mesh or equivalent	3	18	12

THE WOODSTOVE

stove. If you go outside, you'll often see a stream of sparks spewing from the chimney.

If such a blaze begins, call the fire department immediately! If you wait to determine whether you need the professionals' help, it could be too late. Try to deprive the flame of oxygen by closing down the damper. If a stove isn't airtight, this may only reduce the intensity of the fire, or it could produce the creosote-flash pounding effect. Should this occur, open the damper enough to produce a constant burn.

If you can't oxygen-starve the fire, use a flare-type extinguisher (the kind specifically designed for chimney fires and available through most

Chimneys should be checked regularly for excessive creosote buildup and be cleaned, if necessary, by a professional or an informed stove owner who uses the right tools. There is no substitute for a chimney brush.

woodstove dealers), which is a safety device that every stove owner should have close at hand.

Don't pour water into your firebox. The rapid cooling of the stove's metal body can cause it to crack. Some old-timers have suggested throwing a large quantity of salt onto the fire. Such a treatment will help extinguish the flames in the firebox, but it will have a limited effect on the chimney fire and could corrode the body of the stove.

The fire will probably last only a few minutes, but on rare occasions a chimney might blaze for an hour. Keep checking the stovepipe temperature, and watch nearby flammable objects to see that they don't become dangerously hot. In addition, still keep an eye outside to be sure that the sparks coming from the chimney haven't ignited anything in the area, especially your roof.

Once the fire's out, clean and inspect the stove, chimney, and stovepipe, and resolve to do the job more often in the future. The old saying that chimney fires clean out creosote is simply bunk. Often the fire merely transforms the creosote into a thicker, crustier layer, usually leading to very rapid additional accumulations. Check the flue liner carefully for damage, too. Chimney fires can produce temperatures up to 2500°F, which can crack masonry and warp steel.

Despite the fact that woodstoves can be dangerous (as can any appliance, whether it operates on electricity, gas, oil, kerosene, or wood), insurance company statistics show that the vast majority of woodstove-related accidents are the result of either improper installation or lack of maintenance. With the right kind of care and upkeep, your wood heater or cookstove can be a practical, aesthetic, and downright heartwarming addition to your home.

FUEL FROM THE FOREST

A Homebuilt Woodstove

"Water heater" woodburners produce a large amount of heat and a small amount of ashes, and the units hold a fire overnight as efficiently as do many $500 woodstoves on the market.

Most homebuilt woodburning stoves are recycled 55-gallon drums. They more or less serve their purpose, despite the fact that they rapidly burn through and are inefficient, difficult to regulate, and so unsightly that people tolerate them only in the workshop or garage. Low cost is the one redeeming feature of the drum heater. A better way to assemble a homemade woodburner is to use a discarded electric water heater tank, and there are four excellent reasons:

In the first place, the walls of such a tank are at least three times as thick as the metal in a 55-gallon barrel, which means that a water heater drum will make a tougher, longer lasting stove.

Second, when you build a firebox from a junked water heater tank, it's very easy to make the stove as airtight and efficient as many $500 woodburners on the market.

Third, proper construction of the heater means it will be easy to load, it will have excellent fire and temperature control, and it will be presentable in any room of the house.

And fourth, you can build a "water heater" stove for even less than most folks now spend putting

BEST WAY TO LIGHT STOVE: [1] Fill with a couple of pieces of wood on top of a small amount of kindling. [2] Light a wad of paper in "back" of firebox—under flue—to start draw. [3] Then light kindling in front of firebox.

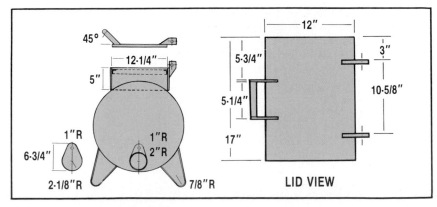

LID VIEW

THE WOODSTOVE

together a 55-gallon-barrel woodburner. If you can find all of the materials as salvage, the stove may cost you only a day's labor to build it. In fact, once you've found your junked, but still intact, water heater tank, about three-quarters of your stove is already "custom made".

Finding one of these tanks isn't difficult because they abound in most junkyards around the country. Any discarded electric, ungalvanized water heater (gas ones are unsuitable for this project) with a 30- to 50-gallon capacity will convert nicely into a stove, but many people feel that the 30-gallon tanks (with a diameter of 20 inches and a length of 32 inches) make the most attractive woodburners.

Pick and choose from your friendly local junk depositories—or from the alleyways behind appliance stores—until you find just the tank or tanks you want. If you're doing your "shopping" in a junkyard, strip off the lightweight sheet metal "wrapper" and the insulation right there, in order to make sure that the main tank inside isn't rusted out or filled with corrosion, excessive mineral deposits, or dirt.

Anyone with a cutting torch and welder will find the rest easy. And if you don't own or operate such equipment, scout around for a competent welding shop that'll convert your tank at a reasonable price.

Lay the container on its side and add legs and the "loading hopper box with hinged lid" as illustrated in the accompanying drawing. Then weld in the "exhaust stack" or "smoke boot" as shown. Make sure that all seams are airtight, and that the hopper box lid fits snugly and is airtight, too. The draft control is, perhaps, the most critical part of all. If it's well made and doesn't leak, you'll have

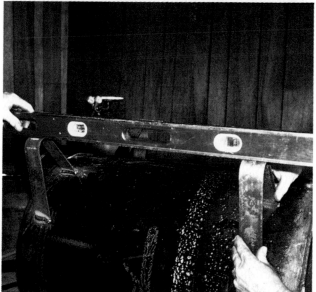

A tank from a discarded electric water heater with a 30- to 50-gallon capacity will convert nicely into a stove. Lay the container on its side and add legs, a loading hopper box with a hinged lid, and a smokestack.

187

FUEL FROM THE FOREST

A Homebuilt Woodstove (continued)

good and positive command of your finished stove's blaze and temperature at all times. So, work carefully and do the job well.

Once the stove is completely assembled, paint all its outside surfaces with rust-resistant black paint or "high temperature engine paint". After proper venting and careful positioning on a heat resistant floor pad away from the walls, the waste-not woodburner is ready for use! And even if you bought everything (approximately 65 pounds of steel) except the recycled water heater tank, the project remains cost-effective.

One of these contrivances will warm an entire 1,100 square-foot house while leaving only a small amount of ashes. Also, the heater can hold a fire overnight. In the morning, just jar the stove a couple of times, open the draft a bit, and the log-burner will snap to life.

You can even cook on this stove, but the appliance's 3/8"-thick top requires such a long heating-up period that the air in the room grows unbearably warm except in sub-zero weather. Before resigning yourself, though, to using a conventional range in all but the dead of winter or escaping to a cooler room to eat the food cooked on the woodburner, try a minor modification.

Get a piece of 1/8" sheet steel that's the same size as the stove's loading door. Then fine-trim this extra "lid" to fit just right and add a handle, and your "cooktop" is in business.

With the thicker top turned back, the thinner plate will warm up rapidly and reach "frying temperature" without overheating the house. It's also capable of several very convenient tricks. For example, the plate (which rests at a slight incline on the loading box frame) can slide back toward the stovepipe to open a crack at the front of the

loading hole for additional air intake. Admittedly, doing so will reduce the fuel's burn time, but the really fast blaze will speed up the early morning coffee.

Then again, when it's turned sideways, the cooktop allows space to insert a popcorn popper or a grill and still keeps the smoke headed up the chimney where it belongs. In fact, the ventilation is so good that you hardly get a chance to smell those T-bones cooking!

By installing a small drum oven, you can even bake with this woodburning heater. Just install the unit above a stovepipe damper, and you'll find that it's nearly always ready for some type of cooking. A banked fire will provide low heat for slowly roasting peanuts, while the blaze left from finishing a meal on the cooktop will leave the oven hot and ready to receive a pan of biscuits.

The stove's air intake, the damper, and the choice of two cooktops make baking control fairly easy. However, the oven is small, and the temperature inside can change rapidly. Therefore, it's best to stay snuggled up to the stove with mending or reading when any delicate treat, such as a cake or meringue, is in progress.

Installing the little oven *will* cut down on the draft to a small degree, but that reduction of airflow will cause trouble only on those days when the stove's draw is particularly sluggish anyway.

Overall, the oven actually contributes to the heating as well as the cooking capacity of the stove. And when food isn't baking, you can leave the door open to enhance this contribution.

Needless to say, life with this little woodburner can be a real joy. The homey stove will heat your house, dry soggy boots and mittens, and cook rib-sticking meals, while crackling cheerfully!

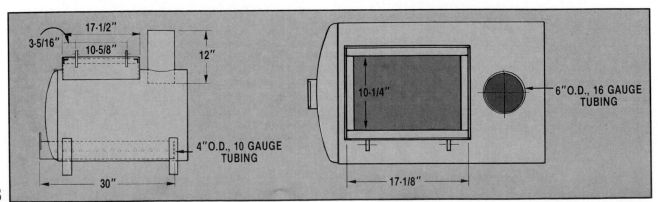

The conversion from a water heater to a woodstove will be a simple job for anyone with a cutting torch and welder. If you don't own or operate such equipment, a welding shop can convert your tank to a stove for a reasonable price.

FUEL FROM THE FOREST

Cook and Heat with a Water Tank?

The kitchen, with its warmth, activities, and mealtime aromas, once served as the center of the home, and the woodburning cookstove was the center of the kitchen. Today's electric and gas appliances admittedly offer greater cooking convenience than did their homey predecessors, but for reasons of aesthetics and independence, the old-timey ranges are enjoying a renaissance. Latching onto a good one, new or old, can cost dearly, however, so why not try building a stove that can inexpensively emulate the original woodburners? This stove of Scandinavian design can provide both an oven and griddle, effective "local" heating, hot water, and a place to dry enough fuel for a day.

The attractive multipurpose cooker shown here is a combination of three nongalvanized scrap water heater tanks . . . one 20″ and the others 22″ in diameter. First, laterally cut the top portions from the two larger vessels, then rejoin those upper sections, top-opposite-top, with a 1/4″-plate cooking surface welded between. To prevent flue gases from entering the baking chamber (and to provide an impediment to smoke on its way out of the stack, thus retaining heat below and behind the oven where it's needed most), fit a 16-gauge contoured baffle wall to the rear of the baking kiln. Allow it to extend about 4 inches below the griddle plate, thereby creating a crescent-shaped passageway that permits the exhaust gases to exit between the stove's outer wall and the oven's rear bulkhead.

The entire stove capsule rests on a 23″-high pedestal cut from the third tank. This stand not only brings the oven to a practical working level, but also serves—with an ellipse removed from its face, and a retaining screen tack-welded in

place—as a storage/drying bin for split wood or kindling.

Many may question the prudence of placing such combustibles near a heat source, but testing has indicated that with the oven operating at a steady 350°F and the ambient temperature in the mid-eighties, the storage area experienced a thermal increase of only 15° above the temperature of the surrounding air. This is sufficient to drive a bit of moisture from wood, but about 100 Fahrenheit degrees shy of actually igniting it. To be on the safe side, however, you should make tests of your own—under the conditions in which you're most accustomed to operating your

This attractive woodburner, which was made from three discarded water heaters, can provide both an oven and a griddle, effective "local" heating, hot water, and a place to dry enough fuel for a day.

190

THE WOODSTOVE

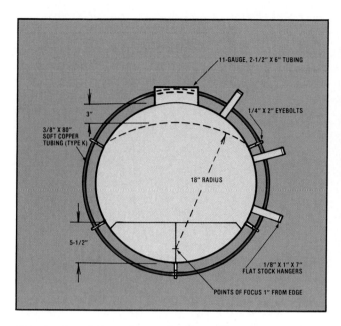

11-GAUGE, 2-1/2" X 6" TUBING

1/4" X 2" EYEBOLTS

3/8" X 80"
SOFT COPPER
TUBING (TYPE K)

3"

18" RADIUS

5-1/2"

1/8" X 1" X 7"
FLAT STOCK HANGERS

POINTS OF FOCUS 1" FROM EDGE

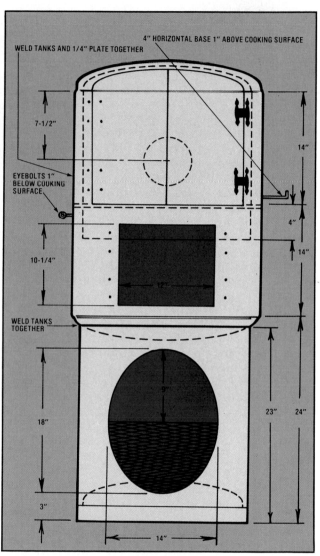

WELD TANKS AND 1/4" PLATE TOGETHER

4" HORIZONTAL BASE 1" ABOVE COOKING SURFACE

7-1/2"

14"

EYEBOLTS 1"
BELOW COOKING
SURFACE

4"

14"

10-1/4"

12"

WELD TANKS
TOGETHER

9"

18"

23"   24"

3"

14"

stove—to determine whether the temperature will remain within accepted limits over an extended period. If any doubt still exists, play it safe by storing your wood elsewhere.

To help prevent warping and joint fatigue in the firebox area and to provide enough thermal mass to absorb and store warmth effectively, line the base and walls of the burning chamber with a castable refractory. Any premixed refractory product should function well.

Protecting the firebox's bottom surface is simply a matter of pouring the loose ''mud'' into the concave base of the hearth to a depth of about 2 inches at the center. To form the *wall* blocks, cut a 9" X 12" flank section from the unused portion of one 22"-diameter water tank, weld a 1/8" X 1"

FUEL FROM THE FOREST

Cook and Heat
with a Water Tank? (continued)

strip of flat stock onto one side of the contoured form, hand-trim two thin sheet-metal arcs and fasten them to the top and bottom edges of the same piece, tack a second 1/8″ X 1″ strip to the curve's inner face at its lower lip, and use a greased block of wood to serve as the adjustable-for-length fourth wall of the mold. Using this homemade mold guarantees not only that the arc of each homemade firebrick is compatible with the inside of the stove, but also that the offset common to the upper seam of most water heating vessels is considered.

The oven doors are cut, with a saber saw, directly from the walls and top of the stove. Leakage through the small joint gap isn't a problem because the cooking chamber isn't exposed to smoke. The firebox doors, on the other hand, overlap the opening by 1/2″ all around and are cut from the lower section of one of the halved 22″ tanks.

The hinges are 1-1/2″ cabinet pieces, but the sets on the lower door are carefully bent to fit the nonflush surfaces. If you can't find suitable latch sets, make your own with two short sections of 3/4″ flat stock—each one notched to fit behind the head of a carriage bolt that you've filed the square shoulder from—and install a pair of heat dissipating coils to serve as handles. The draft controls are simply two 2-1/4″ tube sections welded into the firebox doors and sealed with circular caps, each of which is fastened with a 1/4″ bolt to a threaded strut inside its vents.

To make the stove as practical as possible, you can go on to fabricate a five-gallon copper water tank that snugly follows the contour of the heater's exterior for maximum heat transference. It rests on three L-shaped brackets and is made from a sheet of 16-ounce (one pound per square foot) dead soft copper.

To put the reservoir together, first cut a square-ended, curved pattern block from a piece of 3/4″ plywood. Trim its outer edge along the path formed by swinging a length of string along a 15″ radius and mark its inner boundary in the same manner, but use a radius 4 inches shorter

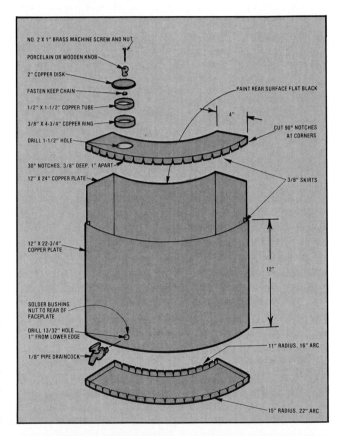

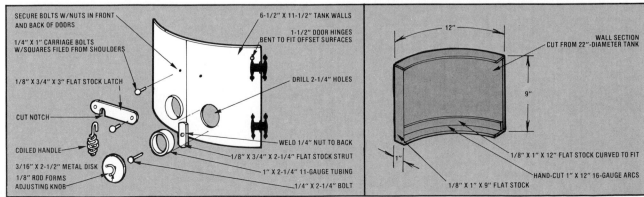

to insure that the inside wall of the tank will rest close to the stove's skin.

Next, cut out two curved plates of copper, each about 3/8″ larger, all around, than is the template block itself, then trim a series of "pie slice" notches—an inch apart and 3/8″ deep—on the arced sides of each plate. (The four remaining straight ends are not trimmed in the same manner, but 90° notches should be cut from their corners to assure a smooth fit after the skirts are folded.)

With this done, slice out two 12″-wide sections of plate—one 22-3/4″ and the other 24″ in length—then fold the skirts on the four edges of the top and bottom sections and the two ends of the tank's faceplate, as in the drawing. Make the appropriate bends, 4 inches in from the side borders, to form the narrow walls of the container.

Before soldering the vessel together, install a spigot and a fill port. The cap is made from a 1/2″ piece of 1-1/2″ copper tube soldered to a 2″ disk, and its receptacle is simply a strip of 3/8″ X 4-3/4″ copper formed into a circle and soldered over a 1-1/2″ hole in the upper surface of the tank. A small knob and keep chain are fastened to the lid with a brass machine screw. The drain-cock has a 1/8″ pipe thread and is held in place by a bushing nut secured to the back side of the container's faceplate.

It's wise to position this drain orifice at least an inch above the bottom of the tank so there's always a bit of liquid left. This assures that the vessel will never get hot enough to melt its solder joints. Since water in this tank may be used for beverages, the solder must be nontoxic.

Finally, to carry the cooker's practicality one step further, surround it with a tubular copper utensil rack that stands away from the firebox and rests in a series of eyebolts. To bend this circular holder to shape, scribe a 25″-diameter circle on a wooden work surface, then tack small blocks of scrap wood at intervals along this ring to serve as a form around which to shape 80″ or so of 3/8″ soft copper tubing. Once it's installed on the stove, the eyebolt fasteners prevent the ring from expanding, so there's no need to secure the free ends of the ring to each other with anything more than a coupling.

After assembling the cookstove, dressing it with several coats of heat-resistant paint, and applying a layer or two of high-temperature flat black to the rear surface of the copper reservoir, load the firebox with five pounds or so of split wood and kindling, put another 25 pounds in the lower storage bin, fill the copper tank with water, and set a blaze in the burner. The results should be impressive: Not quite three-quarters of an hour later, the oven chamber will hold steady at just shy of 400°F, and the stack temperature will be in the same range. The reservoir of water should reach a maximum of 180°, and the refractory mass will absorb enough heat (and subsequently hold it in storage) to tone down fluctuations in temperature which can render a woodstove's baking oven nearly useless.

In its heating mode, the unit does not operate as well as an appliance expressly designed to perform only that function, but it is certainly capable of warming an area the size of the average kitchen, and that's probably as much as anyone would expect. Someone in the market for a woodburning cookstove would do well to give some serious consideration to this design, and the cost plainly isn't prohibitive.

FUEL FROM THE FOREST

Restoring a Cast-iron Classic

Antique woodburning cookstoves and heaters—exquisitely restored to their original elegance—are making a comeback. Not only are they implements that cook food or heat houses, but they are also investments whose values have easily kept pace with inflation.

Unfortunately, all this newfound popularity has caused the price of refurbished stoves to rise dramatically. But just about anyone can own a fine antique without paying a collector's price. The secret? Find a grizzled old woodstove in need of a friend, and restore it yourself!

Perhaps you've never considered refurbishing a stove because you don't know how. But the process, aside from elbow grease, takes only a few common hand tools, some inexpensive supplies, a good electric drill fitted with rotary wire brushes, and—of course—a stove.

Any town over 50 years old probably has an abundance of used stoves rusting away and waiting for someone to adopt them. One of the best ways to find these "experienced" woodburners is to run an ad in a local paper. Antique dealers very often have, or know of, old stoves for sale. Before purchasing a particular stove, though, take a few minutes for inspection to be sure it really *can* be salvaged.

First, make certain the body is sound. Most abandoned woodburners will be rusty, which is all right as long as the metal underneath is still healthy. Potbellied heaters usually were built of cast iron heavy enough to shrug off decades of oxidation, but base burners—self-feeding heaters built of cast iron and sheet metal—might have weak points. Since old kitchen ranges were generally made of lighter material, check their rusty parts with particular care to be sure you

Though looking forlorn in its neglected state, this old woodburning cookstove is still sound and is a good candidate for a restoration project.

have something to restore. Tap any suspicious-looking spots with a screwdriver to make sure they're sound. Inspect the top, bottom, sides, and insides.

Also, when buying a used range, make certain that the oven control (most often a slide or flip-flop damper that directs exhaust heat to the oven) is alive and working. If you don't know what

THE WOODSTOVE

After a great deal of scrubbing, painting, and polishing, the former derelict gleams with new life and is ready again to grace a homestead kitchen.

FUEL FROM THE FOREST

Restoring a Cast-iron Classic (continued)

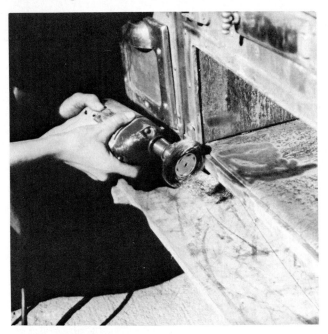

Cleaning corrosion and dirt from a salvaged woodburner can involve a great deal of labor, including chiseling away the scale or whisking off surface rust with a wire brush and an electric drill.

you're looking for, find a stove that is operating properly to use for comparison.

Next, look for cracks. Cast-iron parts are the most likely to be damaged, so scrutinize all cast sections such as the bodies of potbellies and the tops and fireboxes of cookstoves. Hairline fractures may be permissible, but anything bigger presents a problem, especially when the split is in the firebox itself. Don't try to redo a stove for actual use if it has even a tiny crack in the fire chamber. Though iron can be welded, not many welders do it well, and the first blaze might fracture an improperly repaired stove all over again.

After a prospect has passed the soundness and crack tests, examine the trim. Brightwork should be nickel plated, but if the stove's been sitting outside for a long while, there might be nothing to see but corrosion. You can determine if the trim's salvageable by polishing the worst spot with brass polish and 0-gauge steel wool.

Should the trim on a heater be suffering from terminal oxidation, you can have it replated. The bric-a-brac on a range can also be touched up in this manner, but only if it's removed from the stove. In the event that the ruined trim on that cookstove is welded on—or if the stove obviously won't hold together without it—it's better to find another unit.

In addition, damage to porcelain trim—which you'll find on some woodburners—is often sufficient cause for rejecting the stove. Though the ceramic surface is easy enough to clean, the process for repairing chips is prohibitively expensive. So if you can't live with the condition of the porcelain on a cooker or heater, don't buy it!

Some old woodstoves have "windows" (or, by the time you get to them, open holes). These

THE WOODSTOVE

Woodburners came in many shapes and sizes, but the steps in restoring the antiques from the dingy objects found in barns to shiny functional stoves are essentially the same.

FUEL FROM THE FOREST

Restoring a Cast-iron Classic (continued)

viewing ports were usually covered with isinglass, and if the frames are still intact, small panes can be easily and inexpensively replaced.

Once the woodstove's purchased and hauled home, it's time for the refurbishing to begin:

[1] Remove the nickel (or brass or copper) trim. These shiny metals usually serve as the brightwork on a heater, but in some cases, the trim may also be a working part of the stove and not easily removed. Bolts are apt to be rusted, so use lots of penetrating rust solvent to help loosen stubborn ones. Rivets and bolts that won't submit to this persuasion will yield to a hammer and chisel. It's a good idea to replace both hex-head fasteners *and* rivets with brass bolts, since they're easy to install and look nice with nickel.

[2] Rejuvenate the detached trim. If the earlier polishing test was positive, go ahead and shine the trim with brass polish and 0-gauge steel wool until all the corrosion and dirt are gone. Then give the metal a second rubbing with another bit of polish and a soft cloth.

If the trim is beyond reconditioning, it will need to be replated. Locate a metal plater and arrange for the work to be done. (Most platers prefer that you *not* remove the rust before bringing in your piece for refurbishing.) By the way, if the decorative metal is copper or brass—and if you plan to use the stove for more than just a decorative piece—consider plating the piece with nickel. The harder metal is more resistant to oxidation. Be certain not to settle for chrome plating, because this metal will start turning blue with your first hot fire.

[3] Remove the isinglass windows and their frames. If you need replacement panes, and if your nearby hardware dealer isn't able to locate

Sandblasting a rusty old heater with a fine grit such as that used by monument works can quickly remove most dirt and corrosion.

any, you can order isinglass from many woodstove suppliers.

[4] Remove the rust from the stove itself. There are three types of stove to consider, and each requires a different method. If all the ornamentation can be removed from your stove, it's best to have the surface sandblasted, but not with sand: It's far too coarse. Instead, the blasting should be done with carborundum crystals. Look for this service in the phone book under "monument works" since carborundum is used to polish gravestones. Or find someone under "sandblasting" who uses the finer abrasive. Have only the exterior of heaters blasted, unless it's a cookstove you're rehabilitating. In that case, blast the

THE WOODSTOVE

Stove polish applied with a toothbrush or similar implement can help restore a woodburner's original luster. Two coats and a good buffing are recommended.

oven, too. If you're dead set against spending money, you can attack rust with a coarse rotary wire brush on your electric drill (be sure to protect your eyes), but it's miserable work.

The easiest way to remove rust—without damaging shiny alloy trim—is to have the whole unit dipped in a heated chemical bath by an antique-auto stripper. It will be returned to you spotlessly clean (unless it was rusted worse than you thought, in which case you'll get back something as full of holes as a sieve). There might be a little new corrosion where the chemical didn't dry immediately, but that can be taken off with the wire brush and electric drill.

When the stove has porcelain trim and the ceramic parts won't come off (they usually won't), you'll have to go after rusted metal with that coarse rotary wire brush and electric drill. Porcelain won't tolerate dipping or sandblasting, but it will spruce up nicely with tile cleaner and a wet sponge.

[5] Polish the remaining brightwork with a soft cloth. If there's any lingering dirt, use another wad of 0-gauge steel wool for the first pass.

[6] Paint the oven of your cookstove with paint for stainless steel, available in a spray can from your paint or hardware store.

[7] Paint or polish exterior stove parts, but protect the trim. Any cast-iron cooking surfaces should be polished only. On other areas you may interchange paint and polish as you choose.

Brush on stove polish with a small stiff-bristled brush, covering the entire surface thoroughly. Let the brightener dry for a whole day. Then give the unit a second coat and let that dry for another 24 hours. After the second day, buff the stove, using a fine rotary wire brush on an electric drill, until all the residue is removed. Finish up by rubbing the polish with a soft cloth.

[8] Replace any trim that was removed.

[9] Replace the isinglass windows and their frames. Isinglass, which is made from mica, a soft mineral, is easy to work with. Just cut it to size with a pair of scissors.

[10] Give the woodburner a last touch-up with a soft cloth before standing back to admire your handiwork.

It isn't easy. But when the final result is a gleaming work of art that you've transformed from a rusting hulk, somehow all the labor will seem worthwhile.

FUEL FROM THE FOREST

Keep the Fire Where It Belongs

Every well-informed woodstove owner wants to have the safest installation possible. But many folks have discovered that when they follow the recommendations for safe clearances set by the National Fire Protection Association, their heaters end up nearly in the middle of the room. Specifically, the NFPA states that no wood- or coal-burning device should be placed less than 36 inches from an unprotected combustible surface. So when three feet of clearance is added to about three feet of stove, and at least a foot of base pan is included (for safe ash removal), a typical wood heater will protrude some seven feet from a wall. That amounts to a great deal of lost living space, and also means that the stove will probably interfere with normal traffic in the room.

There is, however, a set of NFPA stipulations which state that woodburning appliances can be safely placed closer to walls if an appropriate heat shield is placed between the heater and all combustible barriers. In fact, with such a protector in place, it's possible to position a stove a mere 12 inches from a burnable partition.

By positioning a sheet-metal guard one inch from a combustible wall, it's possible to significantly reduce the temperature to which the wall's surface would otherwise be subjected by heat radiating from a woodstove. The metal barrier is heated by the stove and establishes a rising convective air flow in the space behind it, effectively cooling the back of the steel and preventing the face of the wall from becoming dangerously overheated.

Building a sheet-metal thermal barrier is easy, but there are a few important matters to keep in mind during the designing stage. First, there must be a one- to two-inch space between the floor and the steel to allow air to pass underneath the barrier. And the lumber that braces the metal must be positioned vertically to avoid impeding the convective flow.

In addition, the wooden supports should be insulated to prevent heat transfer from the steel. This can be done by nailing the wood strips to the wall, deep-sinking each nail a quarter-inch into the lumber, and then slipping a section of aluminum window molding between the steel and the wood before tacking the metal sheet in place with 1-1/4" aluminum roofing nails. Because aluminum dissipates heat rapidly, it'll allow very little temperature rise in the wooden stock.

THE WOODSTOVE

Constructing the set of barriers displayed in the photo on this page should take roughly two hours. The job requires a 4′ X 12′ sheet of 24-gauge galvanized steel, 24 feet of window molding, three 1-1/16″ X 1-1/16″ X 8′ boards, twenty 1-1/4″ roofing nails, and a can of flat-black high-temperature paint. (In this case, part of the steel sheet can be used to underlie the brick base upon which the stove sits.) The heater stands 16 inches from the walls, and the wood behind the barriers should remain barely warm to the touch, even when the stove is piping hot.

Another, perhaps simpler, answer to the wood-stove heat dissipation problem can be found at a local heating supply store. Buy a 4′ by 10′ sheet of foil-covered fiberglass duct board, and slice it to match the dimensions of the wall area you're going to protect. (See the photo on page 200.)

After the thermal barrier's been cut to size, it's a simple matter to locate the studs in the wall and secure the panels firmly—foil side out—using 1-1/4″ aluminum roofing nails. (The exposed edges can be trimmed with aluminum tape for a neater appearance.) Finish the face of the boards with heat-resistant paint, and the project's complete. The insulative material will absorb heat evenly over its entire surface, and providing the stove is positioned a commonsense 18 inches or so from the wall, the wood behind the shield will remain protected . . . at an easy-to-live-with cost.

A barrier that protects combustible surfaces from the heat of a woodstove, whether the safeguard is foil-faced and applied directly to the wall or one of sheet steel that is mounted to allow air to circulate behind it, will permit the stove to be installed within 12 inches of the wall.

FUEL FROM THE FOREST

Where There's Fire, There's Smoke

Ever since the first alarms about woodstove emissions were sounded, there's been a flurry of research done on the subject. Findings by more than a dozen different laboratories have confirmed that "airtight" woodstoves do emit large amounts of carbon monoxide, particulates, and unburned hydrocarbons.

Perhaps the newest and least understood of the woodburning-related pollution problems recently came to light when it was discovered that many woodstove-equipped homes have *indoor* levels of carbon monoxide, breathable particulates, and certain members of the POM (polycyclic organic matter) family that are more than ten times greater than *outdoor* measurements taken at the same time. Unfortunately, it'll be a while before we learn just how such a predicament occurs.

There is, of course, a positive side to the recent interest in woodstove emissions research: We've learned more about the physics of combustion in the last couple of years than in all the centuries since humankind discovered fire. In turn, new stove designs and aftermarket products are rapidly being developed to deal with already recognized pollution problems. In addition, a wealth of information concerning how woodstove owners can operate their stoves in a cleaner fashion has recently become available.

And there are fringe benefits to the pollution studies, as well. Creosote—the scourge of every woodburner—is produced under the same conditions as are pollutants. The gooey substance that clings to stovepipes is created by unburned material expelled from the fire below. Though the actual ratio of creosote deposits to emissions varies depending on such factors as flue gas velocity and stack temperature, more pollution is almost always accompanied by heavier creosote formation. An average stove installation leaves about 10% of its particulate and hydrocarbon emissions in its flue, in the form of creosote.

What's more, most schemes devised to reduce emissions and/or creosote buildup tend to improve the overall efficiency of woodstoves. Since most of the materials that make up smoke are combustible, a popular approach to preventing the offenders from reaching the atmosphere is to burn them, either in or above the fire. Doing so also increases the amount of heat that can be obtained from a given amount of wood.

Despite the fact that the study of woodstove emissions and efficiency is in its infancy (and therefore, that various groups use different test methods and arrive at sometimes conflicting conclusions), researchers are unanimous about one facet of the pollution/creosote production process: Every study has found that operator methods can play a larger role in reducing creosote and emissions than can any other factor that's yet been examined. As a result, a clear-cut set of instructions for clean burning has been developed.

The five rules that follow are mostly matters of common sense. But in order to understand why they work, it's necessary to understand a little about how wood burns. The combustion process has been theoretically divided into three phases: evaporation, where the moisture in the wood is removed ... pyrolysis, which is the release of volatile gases trapped in the fuel's structure ... and charring, during which the material's carbon (in the form of charcoal) is burned. However, as

THE WOODSTOVE

systematic and neat as this outline sounds, it is complicated somewhat by the fact that the different stages almost always overlap. Only at the very beginning and end of a burning cycle (early evaporation and final charring, respectively) are the distinctions clear.

The emission of particulates, hydrocarbons, and carbon monoxide is primarily the result of two different situations that can arise during the phases of burning. In the pyrolysis stage, either a lack of oxygen or inadequate temperature (it must be 1100°F or more) above the fire will allow the volatile gases to escape without burning. Any of several factors could lead to either of the conditions. For example, if the stove's draft is nearly closed, there may not be enough air to allow complete combustion of the pyrolytic products.

Emissions can also result from incomplete combustion in the charring stage. And, again, the culprits here are too little air and/or heat. Oxygen starvation might result from a closed-down draft or even inadequate separation of the individual pieces of fuel, which can prevent air from reaching the burn area. Heat can also be lost in other ways, one of which is the addition of new wood to a bed of coals. The evaporation of the moisture in the new fuel can draw large amounts of heat from the embers.

The list of factors that could trigger either of the two major pollution-producing combustion situations goes on and on, but the rules of thumb that follow have proved—by actual experimentation—to deal effectually with many of them.

Rule 1: Use the largest-diameter logs that will burn effectively. Big pieces of wood have less surface area per unit of volume, which prevents them from releasing volatiles too rapidly. This has been recognized as the single most effective technique for reducing emissions.

Rule 2: Build as small a fire as is practical. A stuffed firebox often leads to areas of pyrolysis and/or charring that are too cold or cannot be reached by an adequate air supply. Therefore, use as few of the large pieces of wood as you can while producing adequate heat.

Rule 3: Keep the fire hot. Position the logs in your stove so that air can move through the fire zone, and be sure there's sufficient draft opening. Since you're already trying to make the fire as small as possible, you can maintain high temperatures inside the stove without overheating your home.

Rule 4: Don't increase or decrease the draft setting drastically. Pyrolysis continues for some time after the air supply has been cut back, so slamming the damper shut can send much of your hard-won fuel up the chimney. On the other hand, rapid opening of the damper can carry the pyrolytic products away from the fire too quickly, especially if there's a significant wind-induced draft.

Rule 5: Avoid excessively wet or dry wood. Logs that are too dry pyrolyze very quickly, overloading the combustion zone with volatile gases, while very moist timbers can inhibit effective combustion by absorbing heat for evaporation. Standard air-dried soft or hard firewood (with about 20 to 25% moisture content) seems to be the cleanest-burning fuel.

Naturally, in order to observe these five rules, some stove owners will have to change their habits slightly. Heaters will require loading more frequently than was the case during the era of

FUEL FROM THE FOREST

Where There's Fire, There's Smoke (continued)

the all-night burn. But on the positive side, many of us won't split logs as thoroughly as we have done. Also, following these procedures will cut down the pollutants coming from your woodstove, and it will help to keep your stovepipe cleaner and allow you to obtain more heat from a given amount of wood.

One widely explored technique for cutting emissions and increasing efficiency involves encouraging secondary combustion. With this approach, specific provisions are made to foster the burning of pyrolytic products away from the fire. Many modern stoves have secondary draft controls that are designed to introduce combustion air to the secondary zone. If such a combustion actually took place with any regularity, these appliances would burn considerably cleaner than they do. Unfortunately, most researchers have noted that it's extremely difficult to establish and maintain a secondary burn with anything less than "wide open" fires. In fact, among the conventional heaters equipped with secondary air inlets, it's fair to say that such "afterburning" rarely occurs.

As part of another attempt to produce "cleaner" woodburners, scores of stoves with catalytic combusters have appeared in the marketplace during recent years, as manufacturers have attempted to provide consumers with technological improvements. The honeycomb ceramic and noble metal (platinum, for one) devices work by initiating combustion at a lower temperature than would normally be possible without being consumed themselves. A typical converter begins affecting the oxidation of carbon monoxide at around 500°F. Rising temperatures help to ignite heavier combustibles as well. In practical terms, a catalytic device makes secondary combustion occur at temperatures between 300° and 600°F lower than would normally be possible.

However, the high-temperature furnace is probably—so far—the most thoroughly proven low-emission woodburning approach. The original high-turbulence (as it's often called) furnace is a device that achieves excellent combustion efficiency and low emissions by burning wood at an extremely rapid rate in a well-insulated ceramic firebox with forced-air intake. The fuel is quickly consumed, and the heat is absorbed by water for storage. The emission levels of such furnaces are quite low. They are, in fact, comparable to those of burners fueled by oil, though they're significantly lower in sulfur content.

As is the case in the automobile fuel-mileage gimmick business, there are a lot of would-be inventors in the woodburning industry who'd love to come up with something that a person could stick on the outside of an existing stove to reduce emissions and/or creosote. So far, a number of products have appeared which are said to do just that, but accurate testing lags far behind the claims.

There are several items on the market that assist woodstove owners in monitoring the operating temperature of their appliances, and such products can help a woodburner comply with Rule 3 for clean operation: Keep the fire hot. Furthermore, when one has a thermometer to watch, it's an easy matter to keep track of heat output from the comfort of an armchair.

Yet another approach to maintaining consistently high operating temperatures in a woodstove is the thermostatic air intake control. Many commercially available stoves come equipped

THE WOODSTOVE

with such mechanisms, but in general the units respond too slowly to maintain very even stove temperatures. While the lag time in the reaction of the typical bimetal coil causes many thermostatically controlled stoves to oscillate around the desired temperature, there is a new thermostatic intake air regulator that's been designed to react very quickly to changes in stove temperature. Test results indicate that it is able to maintain very even temperatures in woodstoves, and thus to eliminate the inefficient over- and underfiring common to most heaters.

On the outbound side of the process, several different techniques have been developed to encourage proper draft in woodstove flues. There are a number of chimney caps which—according to their manufacturers—increase flue gas velocities and thereby reduce creosote formation. (One such firm claims that its product will cut creosote accumulation by as much as 75%.) Many advertisements also state that chimney caps may improve combustion efficiency, which could—as an indirect effect—also result in lower emission levels. This is still very much in question, however, since an increase in flue gas velocity (which might reduce creosote accumulation) wouldn't guarantee a corresponding drop in emissions.

Barometric draft regulators have also caused a bit of a stir in the marketplace recently. The devices are similar in concept to oil and gas furnace regulators. As flue gas velocity rises, a damper opens to admit air into the chimney, which prevents overburning caused by excessive draft and adds dilution air to the stove's exhaust. Barometric draft regulators do seem to reduce creosote accumulation, but a lowered rate of emissions may not—in this case—go hand in hand with a cleaner chimney.

Also, some companies now sell catalytic combusters as retrofit devices for woodstoves, but at current prices such a unit will add about as much to the cost of a woodstove as you'd expect to pay in supplemental charges for a factory-installed system. Still, these add-ons should reduce emissions and increase the overall efficiency of conventional woodstoves.

In conclusion, it should be said that the best method for lessening woodstove pollution problems, cutting back on creosote, and easing the burden on our nation's forests is to reduce our need for heat. More and more folks are switching to wood fuel for its economic advantages, and in many parts of the country the load on the environment—in the form of pollution and deforestation—has already become critical.

So before you buy a bigger stove or head out to bring in another cord of wood, consider the possibility of weatherizing your house. There's a good chance that adding insulation, sealing cracks, and putting up storm windows may prove to be less expensive—in the long run—than buying an extra grove of trees. And give some thought to installing solar devices as well. Buying or building them will cost dollars and energy, but once they're in place, the sun-gatherers will usually require a minimum of attention for years to come.

The lessons that we're now learning about woodstove pollution are actually a reiteration of the lessons we've already been taught by coal, oil, gas, and uranium: Too much dependence on any nonrenewable energy source will result eventually in its total depletion.

FUEL FROM THE FOREST

A Functional Flue Alarm

The danger of fire is a constant and legitimate concern to everyone who heats with wood, and there are several measures a person can take to minimize the chance of a blaze. Among these—and one of the most important for the woodstove user—is keeping a watchful eye on the temperature of the stovepipe to avoid the possibility of heavy deposits of soot and creosote igniting in a chimney fire that can spread throughout a house.

At least one smoke alarm and one fire extinguisher should be strategically located in every home that has a woodstove. And in fact, such devices can be lifesavers in any dwelling, regardless of how it's heated. A smoke detector can provide a warning even before an actual blaze gets started, and by doing so helps protect a building's occupants from the primary danger to people in fires: smoke inhalation.

But in addition to these standard safety items, woodstove owners should also employ other fire-warning devices. Several companies now market alarms that signal the overheating of a wood-burning appliance's flue and give an early warning of the possible outbreak of a chimney fire. At current prices, commercial units are hardly an unwise or bad investment, but any handyperson who enjoys tinkering (and saving a few dollars at the same time) will find that building a flue alarm can be an easy, enjoyable project.

A homemade device can be fabricated using readily available components that cost about a third the price of commercially built alarms. The alarm is self-contained (its power is drawn from a 9-volt battery), and it signals overheating with a loud buzz.

The thermometer part of the device is modified by adding a pair of automotive ignition points to the dial. One contact is silver-soldered (regular tin-lead solder might not stand up to the heat of the flue) to the temperature-indicating pointer, and the other is attached to an arm (with a small, insulating piece of circuit board between the two components), which pivots under the bolt in the center of the thermometer. The second contact, then, can be set to a specific temperature, and when the pointer touches it, the circuit is completed and triggers the alarm. While all this is going on, of course, the thermometer still serves its original purpose: indicating flue temperature.

The electronic portion of the audible alarm is controlled by an LN 386 integrated circuit chip. The tiny amplifier circuit that the chip controls consists of seven electronic components, which can easily be soldered to a piece of perforated circuit board, all to be housed in a plastic box.

The circuit should also include a normally open, momentary push-button switch, which can be used to test the battery's condition. This is wired across the two plugs leading to the thermometer, so that the circuit is closed—sounding the alarm—whenever the button is pushed.

Assembling the device takes no more than a couple of hours. Start by modifying the thermometer, which involves silver-soldering an automobile ignition contact to the pointer of the device (make sure the new piece doesn't interfere with the indicator's motion) and fabricating the alarm-set arm from a 1/4" X 1" strip of 20-gauge copper.

Then attach the insulating circuit board strip to the arm with a pair of small pop rivets, and secure the alarm-set contact with a little screw and nut. (Be certain that no part of the contact

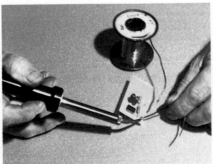

touches the frame of the thermometer, or the alarm will sound constantly.) Next, the positive lead must be silver-soldered to the alarm-set contact, and the ground should be screwed down—along with the alarm-set arm—beneath the central pivot bolt.

To prepare the circuit board, solder all the components to the perfboard, using a low-wattage pencil iron and small-diameter, rosin-core solder. Don't apply heat any longer than necessary when soldering the electronic parts (particularly when securing the IC chip and the capacitors), since excessive temperatures can damage the sensitive components.

With that done, drill two 3/8" holes in one end of the plastic box to accommodate two plug-in

sockets. (In addition, four 3/32" holes are needed for the screws that secure the sockets.) A series of 1/8" noise-outlet holes must also be drilled in the cover (in a circular array), but the speaker itself can simply be glued with silicone sealant to the inside of the cover. Then locate the battery test switch on the side of the box, mounting it in a 5/32" hole.

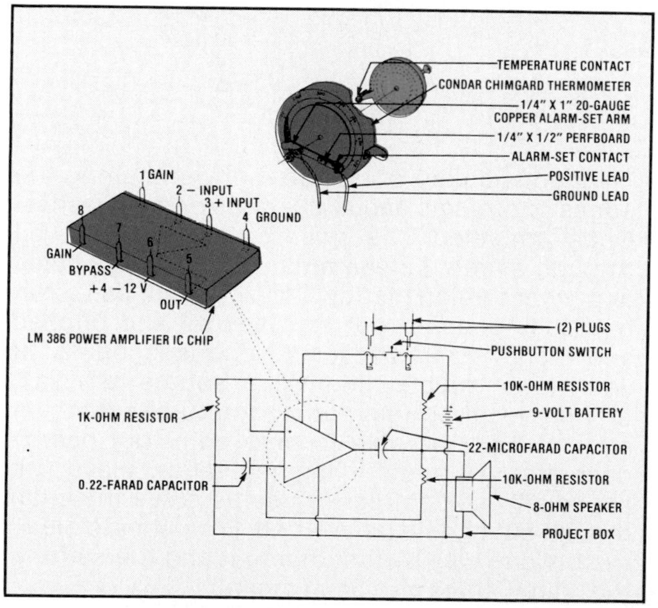

To complete the alarm unit, simply wire the switch, circuit board, battery, and plug-in sockets as shown in the schematic diagram. To keep the components from rattling around, slip the circuit board into the slots provided in the box and separate the battery from the remainder of the parts with two more sections of perfboard.

Since high temperatures might weaken the unit's magnetic hold, you should screw the thermometer to the stovepipe (using tabs that are attached to the device through the holes that originally held the handle). A spot about two feet above the stove is usually a good location. It's also a good idea to install the alarm box some distance from the stovepipe, to protect it and its components from heat. Finally, set the alarm contact at a position which is slightly above the stovepipe's normal maximum operating temperature, light a hot fire, and make sure the device works.

A flue alarm such as this one should provide you and your family with an extra measure of security against chimney fires. But it's no substitute for proper stove maintenance, so inspect the chimney for heavy creosote deposits regularly, have it cleaned as needed, check the batteries in both the flue and smoke alarms periodically, and take every precaution for safe and clean woodburning.

Diagram labels:

- TEMPERATURE CONTACT
- CONDAR CHIMGARD THERMOMETER
- 1/4" X 1" 20-GAUGE COPPER ALARM-SET ARM
- 1/4" X 1/2" PERFBOARD
- ALARM-SET CONTACT
- POSITIVE LEAD
- GROUND LEAD
- 1 GAIN
- 2 – INPUT
- 3 + INPUT
- 4 GROUND
- 8
- GAIN
- 7
- BYPASS
- 6
- 5
- +4 – 12 V
- OUT
- LM 386 POWER AMPLIFIER IC CHIP
- (2) PLUGS
- PUSHBUTTON SWITCH
- 10K-OHM RESISTOR
- 1K-OHM RESISTOR
- 9-VOLT BATTERY
- 22-MICROFARAD CAPACITOR
- 0.22-FARAD CAPACITOR
- 10K-OHM RESISTOR
- 8-OHM SPEAKER
- PROJECT BOX

It takes only a small number of parts and a few construction steps to give your family added protection against a fire.

FUEL FROM THE FOREST

The "Dirt Cheap" Mud Stove

Homemade stoves don't have to be fabricated from iron and steel. Consider, for example, the *louga*, *voltena*, *kaya*, *chula*, and *lorena*. Though these terms may sound like mysterious incantations from a sorcerous ritual, they're all types of simple earthen stoves ... do-it-yourself appliances that can revolutionize cooking and reduce fuel consumption.

Developed, for the most part, by students of appropriate technology working in Third World countries, sand/clay cookstoves are beginning to replace traditional fuel-gobbling open fires in those areas where deforestation, desertification, and dependence on imported fuels have long hindered development.

And, impressively enough, this growing "kitchen revolution" has already decreased wood usage sufficiently to allow many such regions to take the first slow steps toward reforestation: The stoves can, you see, conserve as much as 50% of the wood (or charcoal) that would be required to prepare the same food over an open fire. And, on a more personal scale, earthen ranges provide safer cooking conditions and produce less smoke, lowering the risk of pollutant-related respiratory ailments.

Furthermore, while the initial motivation behind the development of mud stoves was the overwhelming need to ease the serious energy crisis facing Third World countries, the simple-to-build units could certainly be valuable to many people in North America as well. The fact is that anyone who'd like an attractive, inexpensive outdoor stove (to be used, perhaps, for summer canning and baking or even barbecuing) could do a lot worse than to build an earthen cooker, and the needed sand and clay are easily available.

Fine sand can, of course, be found along the edges of streams and at the ends of rain washes. Once collected, the grains should be sieved through a 4mm screen mesh to remove pebbles and other foreign matter. Clays, in turn, can often be easily scouted out in roadcuts and ditches and along streambeds with exposed banks. If you have trouble finding a source of clay, though, local potters and agricultural extension agents might be able to help point out nearby deposits. As you'd imagine, sandy clays are especially well suited for use as stove material, though silt should be avoided. (You should clean freshly dug clay by first drying it and then sifting the "dust" through a 5mm mesh.)

The homemade mud stove is safe, attractive, inexpensive, and fuel-efficient. The earthen cookstoves shown here conserve as much as 50% of the wood or charcoal that would be required to prepare a meal over an open fire.

THE WOODSTOVE

Once a generous quantity of raw material is collected (your own stove design will determine exactly how much you'll need, but do gather three to four times more sand than clay), you're ready to produce a mix with the proper consistency. Appropriate combinations can range—depending upon the specific characteristics of the substances used—from a sand/clay ratio of 2:1 up to one of 5:1.

The ingredients, in dry form, should first be thoroughly mixed and ground underfoot on a clean surface. When the clay appears to be well pulverized into the sand (30 minutes of mixing for a 20-gallon batch isn't unusual), water is added slowly, blended in, and then shovel-mixed.

The most reliable method of testing for the correct sand/clay/water proportion is to make and compare sample bricks from several different mixtures. Use a hammer handle, a length of 2 X 4, or your fist to compact the material into thoroughly wet rectangular wooden frames. Be sure to compress it until your finger can't penetrate the surface deeper than about 1/2 inch, and be especially careful to pack down the mix along the edges and in the corners where the finished bricks are most likely to split or crumble. Then remove the molds and leave the blocks to dry in the sun, remembering to turn them regularly to assure even drying.

To determine the suitability of a clay sample, just wet a handful and work it into a pliable consistency. Then try to mold a thick rope of it around your finger: The mud should retain its workable state and form a ring easily. If it does, you can next flatten the test piece into a rough tile (1/4" or more in thickness), dry it, and "bisque" it in a small fire. After the flames die, leave the clay tile in the hot coals for half an hour before removing it to cool. The finished piece should resist crumbling when scratched.

The success or failure of any particular sample mixture will be readily apparent: If cracks form across a test brick's surface, you've got too much clay (resulting in quick shrinkage), and if crumbling around the edges occurs with reasonably gentle handling, there's too much sand. The ideal blend of the two components will yield a dry block which—although it'll shatter if struck

FUEL FROM THE FOREST

The "Dirt Cheap" Mud Stove (continued)

sharply—has good cohesion and compressive strength.

After determining the best possible recipe and mixing the raw materials, it's time to begin to shape the huge lump of earth into a cookstove. While there are many different designs in use around the world, a good design to begin with is the *lorena* stove, a versatile cooker that was originally developed for use in the highlands of Guatemala and is now being used in several other Central American countries. Its name, by the way, is a composite of the Spanish words for mud (*lodo*) and sand (*arena*).

This particular design allows hot gases from the fire to traverse a tunnel that winds through the stove's body, and then to exit via a simple chimney. While topside cooking is in progress, the lorena's great earthen mass stores heat, which can then be used for extended food warming or for baking, even after the fire has died out. The rate of burn is controlled by adjusting one crude but effective scrap-tin damper at the opening of the firebox and another just in front of the exit flue.

The cooking pots fit tightly into deeply recessed holes on the stove's upper surface. It's best to limit the number of pot holes to no more than three, but their size and shape will depend upon the specific utensils you plan to use. The tunnel must pass under and connect the holes, forming a bend beneath each pot to assure maximum heat transfer.

You'll probably want to sketch an outline of the stove top on the ground or floor next to the construction site in order to determine the dimensions and placement of the pot holes and the position of the flue. Using that drawing as a guide, you can go on to form a solid, level base, either of lorena mixture or of stone or old brick. This base is needed because any shifting of the foundation could cause cracks in the stove's body, and also because it elevates the unit to a convenient height. Next, working upward in straight vertical lines, form the stove itself.

The stove's body will be built up, in layers, to form a solid block into which you'll later carve the pot holes and tunnel. Start by shoveling some of the lorena mixture onto the base and spreading it out to an overall depth of five to six inches. Then compress the material with a homemade wooden tamper (again, a piece of 2 X 4 will do) as you did the test blocks.

Add successive layers in the same manner, taking care always to maintain the stove's vertical line. When tamping near the perimeter of the structure, you'll need to support the outer edge with a wooden frame. Use either right-angled or curved forms at the corners, depending on the shape you've chosen. As you build, make sure all the layers of lorena have the same consistency in order to create a strong, unified mass, and always cover the block with a plastic sheet whenever you have to stop work. (If the material becomes too dry at this point, the interior carving that'll be done later will be quite difficult.)

The finished height of the stove will depend on the depth of its largest pot: Align that container's rim with the cooker's top edge, and then make sure there's a good six to eight inches of solid lorena between the pot's bottom and the stove's base, to allow room for tunneling. When the block finally reaches the proper height, scribe the position of the pots, the tunnels, and the chimney hole on its surface. As you lay out the design, be sure to leave a three- or four-inch margin (about the width of a fist) between the pot holes and the stove's edges.

Just as the lorena stove can be molded from simple materials, it can also be carved with rudimentary tools. In fact, a large metal spoon is all you'll need for scraping out the firebox and the inner passageways. Remember, though, to keep your utensil wet and to leave an insulating layer of lorena (at least two inches thick) between the foundation and the bottom of the firebox and tunnel.

Begin burrowing by cutting into one side of the block to form the firebox. That opening needs to be large enough to accommodate the fuel you'll be using, but—for maximum efficiency—it should be no longer or wider than the first pot hole, which usually is located directly over the firebox.

Next, rough out the pot holes and check the fit of each utensil in its assigned "nest" from time to time. At this stage, however, let the holes stay slightly undersized, since their final forms are best shaped after the rest of the carving is completed. Leave a bit of shoulder to support the bottom of each pot, and then dig about four inches deeper than that (go down six inches or so under the first pot, to make room for the fire) to the

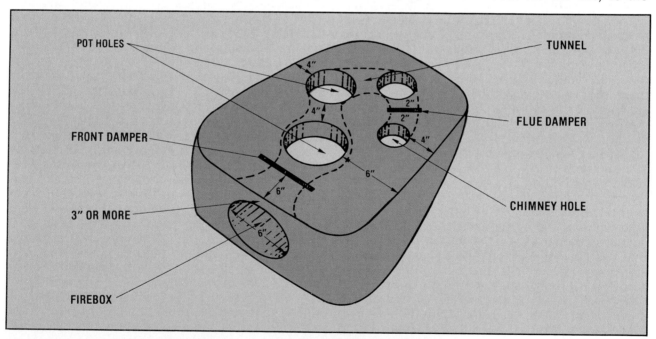

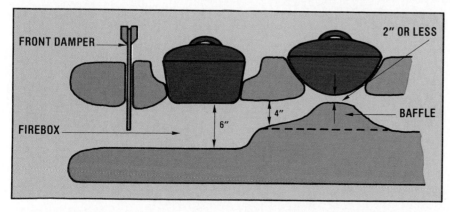

The "Dirt Cheap" Mud Stove (continued)

level that will become the floor of the heat passageway. You can scoop out the chimney hole at this time, too.

Once all the holes are properly spaced along the top of the stove, go on to complete the internal tunneling. Reach through the firebox and the other openings as necessary and dig carefully along your inscribed path, intercepting each pot hole and—finally—the chimney hole. Once it progresses past the firebox, the tunnel should be made barely large enough to accommodate the width of a hand holding three eggs. To institute an even flow of hot gases and smoke, the passage must also be relatively smooth and gently curved.

Following this same excavation process, "fine-tune" the pot holes so that they'll allow the cooking utensils to fit snugly in place. Use the containers themselves as tools (remembering to wet their outer surfaces first) by twisting and easing them into place, but without putting any significant weight on them while doing so. To insure that it gets the maximum exposure to the heat circulating inside the stove, each pot should be recessed until its rim is almost level with the stove surface.

With the digging and shaping completed, it's time to install the dampers. The front closure can simply be attached to the outside of the stove with runners, and a flue damper can be integrated conventionally into the length of stovepipe that will be inserted into the chimney hole. Or slots may be cut into the stove body itself—across the firebox entrance and in front of the chimney hole—to accommodate pieces of thin sheet metal cut to size and bent at the top to form handles. Make the slots by carefully easing

a large knife or machete down into the lorena—using gentle pressure—and working the tool slowly (to prevent cracks) as you form the vertical slits.

The last additions to your homemade cooker will be the baffles, which act as airfoils to direct the flow of hot gas and flames toward the bottoms of the pots. Each baffle—which is actually made from an extra lump of fresh lorena—is formed at the juncture of tunnel and pot hole and rises to within two inches of the pot bottom. Cupping a hand downward, you can easily form the baffle and press the mix in place.

To polish off your handmade stove, you'll need to burnish the block's surface (inside and out) with a smooth rock or the back of a spoon. This finishing process helps to seal the surface against abrasion and spills, but you might want to increase the durability of the lorena even further with a thin coat of oil or paint, or a cement or plaster wash. Any sharp corners should, at this time, be rounded with a wet knife.

Depending on your climate and the stove's location, the hand-formed appliance will take from two to four weeks to cure completely. By setting occasional small fires in the stove during that period, you can hasten the drying process and satisfy your curiosity as to how well your creation will function. Shrinkage should be very slight (thanks to the material's high sand content), but the pots and stovepipe will probably need to be adjusted in their holes now and then as the stove cures. Don't worry about the small surface cracks that'll most likely appear while the lorena is drying: They're not a serious threat.

The sand and clay mixture is, of course, water soluble, so it can easily be reconstituted should

THE WOODSTOVE

you need to repair any serious structural cracks or a damaged edge. Just lightly score around the blemish and dampen the area to accept patching. Large sections should be terraced and then rebuilt in layers. In the unlikely event of a cave-in, the collapsed wall can even be replaced with fresh lorena and retunneled.

On the other hand, a water-soluble stove should be protected from rain. If you've built it outdoors, either provide it with a crude shed roof, or cover the cooled cooker with a plastic tarp after each use.

A lorena stove, when properly made, will be quite functional and durable and should provide reliable service for a number of years. The following general pointers on cooking with mud stoves should provide you with all the basic instruction you need. (Remember, though, that each lorena is unique and can be expected to demand that you learn the "tricks" required for its most efficient operation.)

The pot that sits over—or closest to—the firebox will obviously be the hottest. The second, while still quite warm, is usually better suited to simmering and slow-cooking. The pot farthest from the fire receives only residual heat, so it's generally best used for keeping things warm, such as a supply of water.

Remember that while the stove is in operation, careful adjustment of the dampers will greatly increase the unit's efficiency. And after you're through cooking on the surface, the heat trapped in the lorena's mass can be used for baking. To turn the firebox into an oven, first clear out all the coals, place wrapped food inside, close both dampers, and cover the open pot holes to retain radiant heat.

If built indoors, a lorena stove can be modified to provide thermal storage for home heating by extending the tunnel past the last pot hole and through a second mass of packed clay and sand that will absorb more of the heat left over from cooking. This extension can even be shaped to meet specific needs such as serving as a radiant "hot seat" or a platform for a small warmed bed. Because of the mass involved, the surface temperature of such an accessory will rarely rise above a comfortable level, but that heat will be retained for a good long time. (If you do construct an additional earthen section, though, you'll have to include removable tiles or bricks along its tunnel route to permit access for cleaning out built-up soot, ash, and creosote.)

In order to exploit the absorbed heat even further (after all, why waste "free" energy?), it's possible to route a cast-iron pipe through the accessory mass as you build it (this should *not* intersect the flue tunnel). A small fan can then be placed at one end to pull in room air, which will be warmed as it passes through the tube, and to blow it out the other end.

Though the ideas presented here are only a few of the possible variations on a basic concept, you can see that whether it's used as an outdoor cooker or as an indoor source of heat for a home, garage, or workshop, the earthen stove is a simple invention with profound ramifications. By allowing for increased local self-reliance, requiring only readily gathered materials, and conserving precious fuel, this fine example of soft technology offers hope for the deforested Third World, as well as an alternative to the often wasteful appliances typical of our own industrialized society.

FUEL FROM THE FOREST

The Stone Ranger

This primitive "kitchen range" is often used in survival skills training programs given to missionaries who intend to set up housekeeping among native tribes far from civilization. Therefore, the practical cooker/baker could serve nicely in *any* remote area where supplies are scarce and the most common building materials are stone, earth, wood, and salvageable scrap metal.

The first step is to build a platform, which simply raises the cooking area to a convenient working height. This can be done by driving short poles into the ground and tying crosspieces to the uprights with a generous amount of baling twine. When that's finished, top the framework with branches that will be sturdy enough to provide a base for the heavy stove.

Next, construct the bottom and three sides of a stone or brick box, open at one end, that will later hold the fire. To do so, place a covering of stone on the raised platform and cement the rocks together with clay or mortar, keeping the surface as smooth as possible so it will be easier to clean. Then build up the sides of the firebox as you would a wall and seal all the chinks. To provide a cooking surface, cover about two-thirds of the top of the box (toward the open end) with a flat plate. A cast-iron sheet is best, but you can use steel or even a large, flat stone. Remember, though, that only hard, unlayered rock will do, and the slab must be dried out by heating it slowly. This eliminates the risk of its cracking or even exploding during exposure to intense heat.

Now it's time to build the oven over the portion of the firebox that's not covered by the cooking surface. Place a whole 16-gallon drum inside a 30-gallon drum that's had the front end removed, a section cut out along its length, and a hole cut for the chimney. Cradle the smaller drum within the larger one with rocks and support both containers with stones at the ends. The larger barrel serves as a form for the brick or stone that surrounds the entire oven except for the door and chimney openings. Before reaching the chimney, the smoke and hot gases pass through the space between the two drums, heating the oven.

The oven must be fitted with a door. This job is most easily handled with an oxyacetylene torch, but if necessary, it can be done with hand tools. Cut a square hole in the metal end of the 16-gallon drum, make a cover (each of its dimensions should be about an inch bigger that those of the opening), then— with hinges at the bottom so it can swing out of the way—attach the door to the oven and make or buy a latch for the closure.

The chimney, which consists of a pipe set into a hole in the arch above the oven, must fit tightly and be cemented with clay or mortar. You'll also need to put a damper in the pipe so that the draft, and therefore the amount of heat produced, can be controlled.

Of course, this is just one possible form of simple oven. The Indians of the Southwest, and other people around the world achieve the same results by building a stone beehive-shaped structure, with an opening in the front near the bottom and a smoke hole at the top. This chamber is heated through for a couple of hours, the ashes are raked out, and the stored-up heat in its stones can then be used for baking. This form of oven is a little more awkward to use, but it does work well. Although the Indian version is usually built on the ground and therefore requires a lot of bending over, raising the stove on a platform would eliminate the need for stooping.

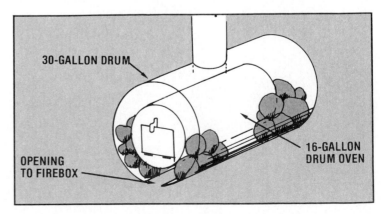

30-GALLON DRUM

16-GALLON DRUM OVEN

OPENING TO FIREBOX

THE WOODSTOVE

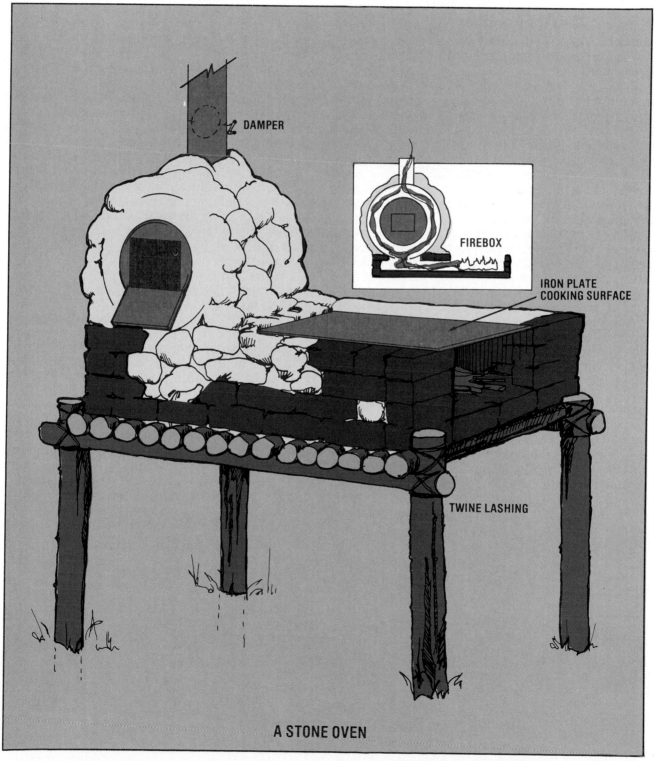

DAMPER

FIREBOX

IRON PLATE
COOKING SURFACE

TWINE LASHING

A STONE OVEN

FUEL FROM THE FOREST

The Magical Aladdin Oven

Back in the late 1800's a fellow named Edward Atkinson devised a highly efficient, inexpensive cooker called the Aladdin oven, and the reasoning behind his invention is just as timely now as it was then.

By his own account, Atkinson became concerned with the economics of contemporary cooking methods after reading some statistics on the cost of living. It seemed to him that a disproportionate part of the household budget was spent on the preparation of food, and he began an investigation that was to last ten years.

Atkinson's studies revealed that the ordinary kitchen practices of the 1800's required the burning of more than two pounds of coal for every pound of food processed. His answer to the high cost of cooking was a device that enabled one pound of kerosene, burned in an ordinary kerosene lamp, to do the work of 50 to 70 pounds of coal. The oven took its name from the heat source: the familiar Aladdin lamp.

The Aladdin oven is really a very simple affair. The cooker itself is merely two-thirds of a common wooden barrel. It's important that the container be made of a material that conducts heat poorly, as this factor governs the oven's effectiveness. The drum is inverted over a sheet-iron table pierced in the center by a hole about three inches in diameter. This opening does not lead directly to the cooking chamber, but is closed off on the inside by a cylindrical deflector made of tin. In this way, the lamp radiates heat into the interior without drying the food or tainting it with the products of combustion.

The heat source—the trusty lamp, properly tended to insure perfect burning—is placed under the table so that the top of the chimney is just inside the deflector's base but doesn't touch the sides of the opening. Food is placed on and around the deflector (or, alternatively, on a raised soapstone plate), and the heat from the lighted lamp cooks the dishes slowly and thoroughly to a state of delicious tenderness.

The general principle behind such a device is quite old. Atkinson's invention has been described as one of the most perfect substitutes for the colonial brick oven (itself a descendant of the brick and stone baking chambers unearthed at Pompeii) and for the Polynesian earth oven. The nineteenth-century inventor's contribution was to adapt the idea to the use of a novel heat source. Apparently Atkinson worked hard on his discovery: He records that over a thousand lamp-powered cookers were made by hand while the design was under development.

Kerosene lantern cookery received a most thorough and practical test on a large scale when it came to the attention of Booker T. Washington, then head of Tuskegee Institute in Alabama. He ordered the construction of a giant Atkinson-type oven made of pine planks, plastered on the inside, and heated by five lamps. This monster was used to feed more than 600 adult students, and the tough local beef and mutton were reportedly rendered tender and delicious by the slow and even processing.

Rather than trying to build an oven exactly to Atkinson's specifications, you might want to construct the version of his device illustrated here, using materials that are readily available. The barrel is easy enough to obtain. Use a table saw to cut away one-third of the cask, then clean the remaining two-thirds by scraping away the char and washing the interior.

THE WOODSTOVE

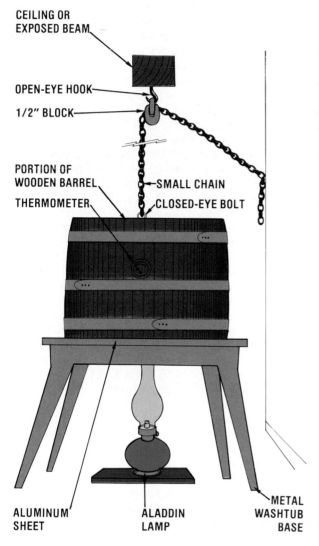

CEILING OR
EXPOSED BEAM

OPEN-EYE HOOK

1/2" BLOCK

PORTION OF
WOODEN BARREL

SMALL CHAIN

THERMOMETER

CLOSED-EYE BOLT

ALUMINUM
SHEET

ALADDIN
LAMP

METAL
WASHTUB
BASE

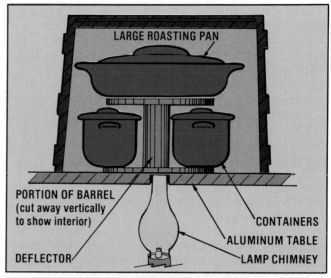

LARGE ROASTING PAN

PORTION OF BARREL
(cut away vertically
to show interior)

CONTAINERS

ALUMINUM TABLE

DEFLECTOR

LAMP CHIMNEY

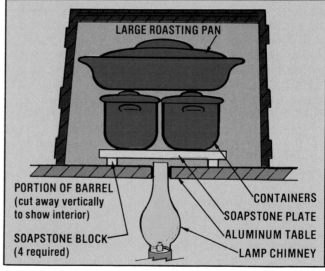

LARGE ROASTING PAN

PORTION OF BARREL
(cut away vertically
to show interior)

SOAPSTONE BLOCK
(4 required)

CONTAINERS

SOAPSTONE PLATE

ALUMINUM TABLE

LAMP CHIMNEY

217

FUEL FROM THE FOREST

The Magical Aladdin Oven (continued)

Next, you'll need some type of base (perhaps a salvaged metal washtub stand) and the tabletop, which can be made from a three-foot-square sheet of iron or aluminum. Cut a hole in the center of the top about three inches in diameter or just large enough to admit the upper end of your kerosene lamp's chimney without touching it.

The deflector can be made of tin or aluminum, and must be designed to fit within the barrel. Besides the cylinder, which should have a diameter of between three and four inches and should be one-half the height of the oven, there are metal support disks on either end of the tube which should measure two-thirds the diameter of the barrel's interior.

Cut the metal for the cylinder to size, making allowance for an overlapping edge when the seam is welded. In addition, make cuts—one-half inch deep and about one-half inch apart—along the top and bottom edges of the cylinder. Then fold these outward at a 90° angle to act as additional support for the upper disk (which will hold a large pot). After cutting a hole in the lower disk that corresponds to the diameter of the cylinder, weld the tube's seam. Then weld the disks to the top and bottom of the tube. To insure a tight seal, the completed deflector should also be welded to the table, and if extra support is needed, braces can be welded or screwed along the edges of the deflector's top and bottom plates. (There should be no more than four braces, and they should not interfere with the placement of the cooking containers.)

Now, though you won't need to be continually checking your food while it's cooking, because the low heat and constant moisture make burning unlikely and stirring unnecessary, you will need a way to raise the oven without touching the barrel, which will become extremely hot. If you securely anchor an eyebolt in the top of the cask, you can then simply fasten a chain to the eye, run the links through a small block attached to the ceiling or an overhead beam, and place a nail or hook in the wall at a height that will allow you to lock the cover in a raised position while you insert or remove pots and pans.

This completes the Aladdin oven, but you might want to "refine" the cooker by installing an oven thermometer such as those once found in the doors of commercial kitchen ranges. Such a gadget will allow you to regulate the heat precisely, use conventional oven-cooking recipes, and produce perfectly cooked meals with little or no experimentation.

As is any piece of large-capacity cooking equipment, the lamp-powered oven is most economical when used to prepare an entire dinner rather than just part of one. Atkinson claimed that a pound of kerosene—a fraction over one pint—would process 20 to 30 pounds of food, *providing* the oven is loaded to capacity. It makes no sense to heat the contraption merely to cook a hamburger or to bake a potato.

For the best-tasting results possible from an Aladdin oven-cooked meal, it's important to determine whether the deflector is airtight. Since this device, when properly installed, keeps the oven's atmosphere moist rather than dry, many dishes need not be covered at all. Meat, fish, and fowl can be placed in any kind of heat-proof container—along with a few tablespoonfuls of water —and cooked with or without lids. Only foods containing a large amount of liquid (cereals, soups, stews, and such) really need covering.

THE WOODSTOVE

However, if the deflector is not completely air-tight, you'll want to be more careful about the containers you use. Clay casseroles, for instance, hold heat effectively but should have tightly fitting covers. Glass and metal pots are generally manufactured with a closer fit between the lid and the vessel, and this prevents the escape of the natural juices and flavors so important to good cooking. An advantage of cookware made of a heat-resistant glass-ceramic combination is that it allows the food to be cooked, served, and then the leftovers stored, all in the same container.

Since the longest-cooking item on a menu is usually the meat dish, it should usually be placed on top of the deflector. Then, when the foods that need less processing are added, the oven need only be raised enough to permit the various containers to be arranged around the bottom of the tin cylinder. Since heat rises and will remain trapped in the deep cover, very little warmth will be lost during this operation.

Speaking of heat, you may be wondering whether a wooden vessel can safely contain temperatures high enough to roast food. Well, the same thought occurred to Atkinson, whose reports indicate he brought the internal temperature of one of his creations to 500°F and maintained that level for eight hours to see whether scorching would take place inside the barrel. It didn't.

In practice, of course, no cooking should be done at such a temperature. It's the long, slow processing of food that produces the finest results, with the majority of dishes baked at between 200° and 250°F.

It's doubtful that scorching will take place if cooking temperatures are kept below 300°F, especially if the airtight metal deflector is used. The device, as opposed to the alternative open soapstone deflector, promotes a moist atmosphere which should retard any tendency the wooden barrel might have to to catch fire. (Nevertheless, the oven should be used on a hearth or in some other place where a blaze could be contained if it *did* somehow start.)

Finally, fireproof or not, the Aladdin oven *is* a peculiar looking object, rather like a practical joke on a French chef. But a joke it surely is not. The idea was born at a time when fuel was, in some cases, even more precious than it is today, and Atkinson's solution is as practical now as it was in 1890!

FUEL FROM THE FOREST

Wood to Oil to Heat
The Triple Play

Though there are many ways to produce heat efficiently and inexpensively, thermal energy is a fleeting thing, and storing the warmth effectively is often difficult and costly. However, by combining a heat source with thermal mass in a single unit, warmth can be produced and saved in an affordable, compact manner. This can be done in a homemade woodstove that incorporates in its design an oil storage and circulating system which gathers and holds those elusive Btu, allowing them to be used over an extended period of time.

This heating device is little more than a woodstove surrounded by an oil-filled chamber which is specifically designed to expose a maximum amount of surface area to the heat source. This optimum thermal transfer is accomplished by encircling the cylinder-shaped firebox with the viscous liquid—a commercial oil formulated to handle high temperatures—and by cleverly routing the stove's smoke (and hence its normally wasted heat) directly through the oil reservoir via a series of 1-1/4"-diameter tubes.

Oil, rather than free-for-the-taking water, is used as a thermal storage medium for several good reasons: [1] Oil—like water—can hold heat for an extended period of time. [2] The boiling temperature of oil is more than twice that of water. Therefore a lot more heat can be fed into it, at atmospheric pressure, before a molecular change occurs. [3] Oil is noncorrosive, and thus its use results in a longer life span for the furnace *and* for the plumbing connected to it.

Another important point to remember (for reasons of both safety and legality) is that, although this oil-heating furnace may look like a boiler, it is definitely not one since it's open to the air and operates at atmospheric pressure. This means—in essence—that those laws, inspections, and other requirements that apply to steam boilers do not pertain to this heater. Also, because any water within the oil will be vaporized—and vented—long before the heat storage medium reaches its operating temperature, neither condensation nor the accidental introduction of water into the oil reservoir will pose a problem.

Naturally, the furnace described here could be used—in conjunction with a heat exchanger—for space heating, water heating, evaporation-based cooling or refrigeration, or for any of a number of other purposes.

The drawings are probably all but self-explanatory, but there are a few tips that will make the task of fabricating this piece of equipment somewhat easier. In most cases, both of the furnace's "drums" will have to be formed from flat plates at your local steel supply house unless two tanks of about the required sizes can be located.

It's also a good idea to have a sheet metal shop stamp the holes in the end plates that seal the extremities of the oil chamber. These two plates must be drilled (or stamped) simultaneously to insure that the perforations line up exactly, and this is a difficult procedure for most amateur metalworkers. In addition, the chimney stack may be awkward to piece together, since the fitting involves joining a circular collar to a four-sided opening, so this component, too, might best be made by a professional.

The furnace design has proved sound, and such a unit has shown to be impressively efficient. Flue temperatures average below 400°F, which means that a good deal of heat is going into the oil rather than up the chimney. And the apparatus is perfectly safe as long as the welds are intact and the oil's

THE WOODSTOVE

This compact woodburning furnace
incorporates thermal storage
capabilities and inexpensive
operation.

FUEL FROM THE FOREST

Wood to Oil to Heat
The Triple Play (continued)

temperature remains below the liquid's flash point of about 435°F.

Because of the heating unit's very affordable cost and the fact that it can burn minimal amounts of fuel and still provide useful energy, this homemade furnace might just solve a lot of people's winter warming problems.

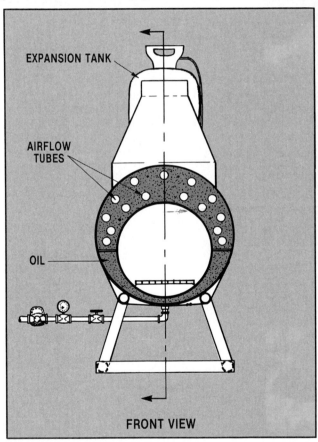

FRONT VIEW

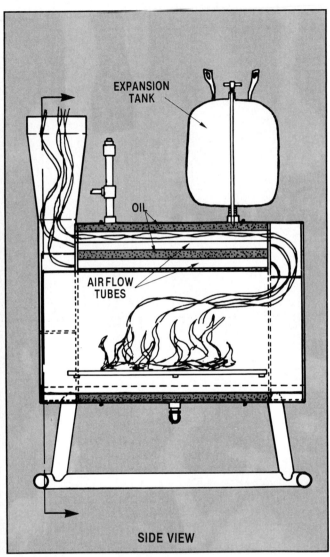

SIDE VIEW

THE WOODSTOVE

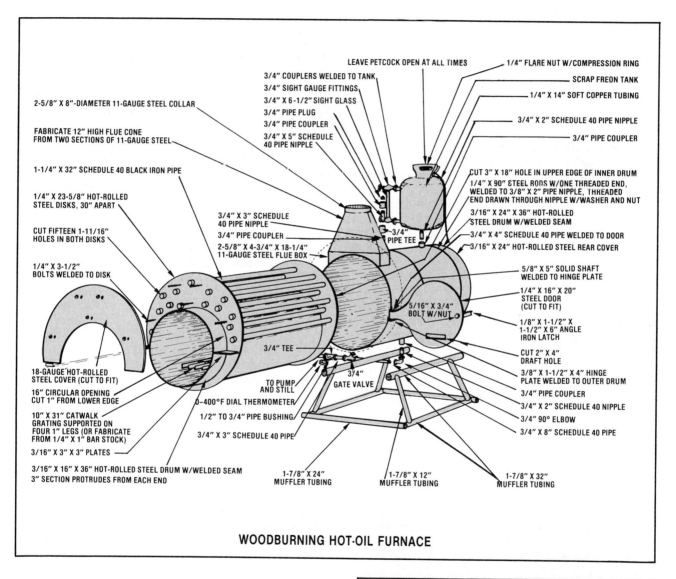

Labels (clockwise/around diagram):

LEAVE PETCOCK OPEN AT ALL TIMES

1/4" FLARE NUT W/COMPRESSION RING

3/4" COUPLERS WELDED TO TANK

SCRAP FREON TANK

3/4" SIGHT GAUGE FITTINGS

1/4" X 14" SOFT COPPER TUBING

3/4" X 6-1/2" SIGHT GLASS

3/4" PIPE PLUG

3/4" X 2" SCHEDULE 40 PIPE NIPPLE

3/4" PIPE COUPLER

3/4" PIPE COUPLER

3/4" X 5" SCHEDULE 40 PIPE NIPPLE

2-5/8" X 8"-DIAMETER 11-GAUGE STEEL COLLAR

FABRICATE 12" HIGH FLUE CONE FROM TWO SECTIONS OF 11-GAUGE STEEL

CUT 3" X 18" HOLE IN UPPER EDGE OF INNER DRUM

1-1/4" X 32" SCHEDULE 40 BLACK IRON PIPE

1/4" X 90" STEEL RODS W/ONE THREADED END, WELDED TO 3/8" X 2" PIPE NIPPLE, THREADED END DRAWN THROUGH NIPPLE W/WASHER AND NUT

1/4" X 23-5/8" HOT-ROLLED STEEL DISKS, 30" APART

3/16" X 24" X 36" HOT-ROLLED STEEL DRUM W/WELDED SEAM

3/4" X 3" SCHEDULE 40 PIPE NIPPLE

3/4" X 4" SCHEDULE 40 PIPE WELDED TO DOOR

CUT FIFTEEN 1-11/16" HOLES IN BOTH DISKS

3/4" PIPE COUPLER

3/16" X 24" HOT-ROLLED STEEL REAR COVER

1/4" X 3-1/2" BOLTS WELDED TO DISK

2-5/8" X 4-3/4" X 18-1/4" 11-GAUGE STEEL FLUE BOX

3/4" PIPE TEE

5/8" X 5" SOLID SHAFT WELDED TO HINGE PLATE

1/4" X 16" X 20" STEEL DOOR (CUT TO FIT)

5/16" X 3/4" BOLT W/NUT

1/8" X 1-1/2" X 1-1/2" X 6" ANGLE IRON LATCH

CUT 2" X 4" DRAFT HOLE

3/4" TEE

18-GAUGE HOT-ROLLED STEEL COVER (CUT TO FIT)

3/8" X 1-1/2" X 4" HINGE PLATE WELDED TO OUTER DRUM

16" CIRCULAR OPENING CUT 1" FROM LOWER EDGE

3/4" PIPE COUPLER

TO PUMP AND STILL

3/4" GATE VALVE

3/4" X 2" SCHEDULE 40 NIPPLE

10" X 31" CATWALK GRATING SUPPORTED ON FOUR 1" LEGS (OR FABRICATE FROM 1/4" X 1" BAR STOCK)

0–400°F DIAL THERMOMETER

3/4" 90° ELBOW

3/16" X 3" X 3" PLATES

1/2" TO 3/4" PIPE BUSHING

3/4" X 8" SCHEDULE 40 PIPE

3/16" X 16" X 36" HOT-ROLLED STEEL DRUM W/WELDED SEAM 3" SECTION PROTRUDES FROM EACH END

3/4" X 3" SCHEDULE 40 PIPE

1-7/8" X 24" MUFFLER TUBING

1-7/8" X 12" MUFFLER TUBING

1-7/8" X 32" MUFFLER TUBING

WOODBURNING HOT-OIL FURNACE

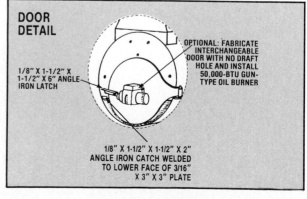

DOOR DETAIL

1/8" X 1-1/2" X 1-1/2" X 6" ANGLE IRON LATCH

OPTIONAL: FABRICATE INTERCHANGEABLE DOOR WITH NO DRAFT HOLE AND INSTALL 50,000-BTU GUN-TYPE OIL BURNER

1/8" X 1-1/2" X 1-1/2" X 2" ANGLE IRON CATCH WELDED TO LOWER FACE OF 3/16" X 3" X 3" PLATE

FUEL FROM THE FOREST

A Tree-Powered Truck!

Model rockets may have the option of getting their boost from either solid or liquid fuels, but cars and trucks seem not to have such a choice. Well, an alternative power system does exist, and it can move a rig down the road as smoothly and reliably as can any conventional source of power . . . and can do so at zero fuel cost!

Here's how the system works: Wood scraps (size is very important, and the chunks used are larger than sawdust or shavings but smaller than a 6″ length of 2 X 4) are contained in a modified hot water tank and rest on a cone-shaped, cast-refractory hearth. The recycled vessel is airtight, except for a springloaded, sealed fill-lid, a capped lighting aperture, and an inlet port (the last is simply a 2″ brass swing check valve, which allows the "draw" created by the engine to pull controlled amounts of air into the firebox).

Incoming "atmosphere" is directed through a series of holes drilled into one shoulder of a discarded wheel rim (which is girdled with a circular band of strap metal and fastened to the bottom of the tank), and supports combustion in the vicinity of the hearth. As the fuel in that area burns, it consumes the oxygen in the air, creating carbon dioxide and water vapor and forming a bed of glowing charcoal that collects on a grate suspended from chains several inches below the hearth assembly. (Simultaneously, a heat-induced "decomposition" zone is created right above the combustion region, driving gases from the wood and carbonizing it prior to incineration.)

The mixture of CO_2 and moisture (with the addition of some creosote) is then drawn through a "choker" (positioned between the hearth and the charcoal grate) and forced into the embers at the lower part of the tank before leaving the gasifier.

The choke serves as an air restricter which blends the various vapors and directs them through the glowing coals, where they're reduced to combustible gases: carbon monoxide, hydrogen, and small amounts of methane. The final product also contains a good deal of nitrogen, along with some unconverted CO_2 and traces of tar and ash.

The carbon dioxide and nitrogen are inert, and such nonfuels pose no threat to the powerplant. However, the tar and ash must be removed from the gas, or they may produce deposits and damage the engine. In order to clean the fuel, the "smoke" is first routed through a liquid-cooled "densifier" (a multitubed heat exchanger surrounded by a water jacket and plumbed into a castoff automobile air-conditioning condenser that's mounted in front of the existing radiator), which precipitates moisture and residue from the gas. Then it passes on to a tubular filter that's [1] packed with strands of commercial air conditioning filament, woven transport padding, or a similar material that won't disintegrate, rust, or burn, and [2] equipped with perforated flame traps at its entrance and its exit.

The final strainer catches the remnants of ash and tar in the gaseous fuel, which then travels through a slightly bowed horizontal tube (where most of the little remaining moisture is trapped) and on into the engine.

To allow for the use of either wood gas *or* gasoline, this system utilizes a unique mixing chamber and linkage setup. Fashioned from scrap carburetor parts, a pair of old brackets, some cabinet door hinges, and three clevises, this hookup seems to suit the nature of producer gas to a "T".

WOOD GASIFICATION

Because the vaporous fuel has a fairly low Btu value (and because the amount of usable power contained in the wood "smoke" can be affected by engine speed, load, moisture, and other factors), the proportion of gas to air has to be much greater than that of, say, a propane-powered engine. The driver must be able to adjust the mix *in transit* to obtain a consistent degree of performance under all types of driving conditions, and yet he or she shouldn't have to be constantly manipulating the controls.

Well, this design meets all of those requirements. First, take a 4" length of 1/8" X 2" X 4" tubular steel, and—using the stock gasoline carburetor and manifold as a template—drill fuel passage and mounting holes through its broad upper and lower surfaces. Next, seal the tube's "inboard" end with a piece of scrap metal and trim out a section of 1/4" X 2-1/2" X 4-1/2" plate. Bore and tap 5/16" mounting-stud holes at each corner of this panel, spacing the openings to correspond with the base of a Ford Autolite/Motorcraft 5200 or a Holley 5210 two-barrel carburetor. (These particular units were original equipment on Pintos and Vegas, respectively, and they should be readily available in wrecking yards.)

The next step is to drill 1-1/4" and 1-1/2" holes in the plate to match the carb's primary and secondary throttle bores, then weld the newly fashioned piece of hardware onto the uncovered end of the tubular steel chamber. Now, make a second flange, identical to the first one except for the fact that its corner holes are straight-drilled rather than threaded. Then relocate the PCV vacuum fitting from the manifold to the box in order to eliminate any possibility of its interfering with the new hardware.

Fuel for this wood-gas–powered pickup truck can be found almost anywhere, including lying in brush piles alongside the road.

At this point, the throttle body should be removed from the rest of the carburetor (just at the bottom of the float bowl and about 3/4 inch below the venturis) with a hacksaw, its cut surface filed flat, and four of its internal passageways—the idle screw feed, both idle transfer holes, and the distributor vacuum port—permanently sealed with small ball bearings (lead shot could be used instead) and silicone.

The remainder of this procedure involves threading the four 5/16" X 1-3/4" studs into place, sliding the throttle body over them (shafts upward), slipping 1-1/4" and 1-1/2" flanged sanitary elbows through the appropriate openings in the second (unthreaded) mounting plate, applying some silicone sealant at the flanged terminus of each pipe, and securing the entire assembly with some lock washers and nuts. The unit is then mounted on the manifold—with the gasoline carburetor on top—after installing carb-mounted studs, 2" longer than the original ones, and gaskets.

In this configuration, with the wood-gas supply hose connected to the 1-1/2" fitting and the smaller elbow leading to the air filter chamber, the engine can be made to operate in several modes. Furthermore, switching from one to another is as simple as pulling the cable that governs an uncomplicated but unique selective linkage assembly.

The system is based on an in-line series of three 3/8" clevises and a modified door hinge, all of which pivot on a single 3/4" X 6" bolt that's fastened to an existing mounting hole in the manifold. The extreme left clevis controls the forward (wood gas) throttle plate, the central fastener governs the movement of the rear (fresh air)

A Tree-Powered Truck! (continued)

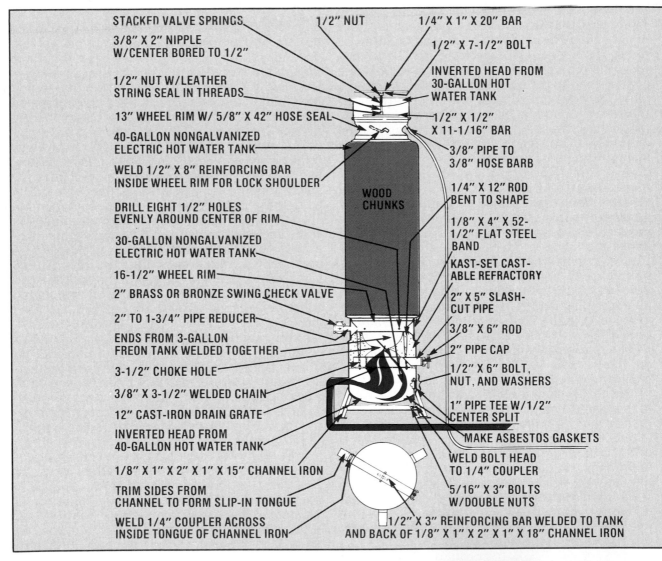

STACKED VALVE SPRINGS

3/8" X 2" NIPPLE W/CENTER BORED TO 1/2"

1/2" NUT W/LEATHER STRING SEAL IN THREADS

13" WHEEL RIM W/ 5/8" X 42" HOSE SEAL

40-GALLON NONGALVANIZED ELECTRIC HOT WATER TANK

WELD 1/2" X 8" REINFORCING BAR INSIDE WHEEL RIM FOR LOCK SHOULDER

DRILL EIGHT 1/2" HOLES EVENLY AROUND CENTER OF RIM

30-GALLON NONGALVANIZED ELECTRIC HOT WATER TANK

16-1/2" WHEEL RIM

2" BRASS OR BRONZE SWING CHECK VALVE

2" TO 1-3/4" PIPE REDUCER

ENDS FROM 3-GALLON FREON TANK WELDED TOGETHER

3-1/2" CHOKE HOLE

3/8" X 3-1/2" WELDED CHAIN

12" CAST-IRON DRAIN GRATE

INVERTED HEAD FROM 40-GALLON HOT WATER TANK

1/8" X 1" X 2" X 1" X 15" CHANNEL IRON

TRIM SIDES FROM CHANNEL TO FORM SLIP-IN TONGUE

WELD 1/4" COUPLER ACROSS INSIDE TONGUE OF CHANNEL IRON

1/2" NUT

WOOD CHUNKS

1/2" X 3" REINFORCING BAR WELDED TO TANK AND BACK OF 1/8" X 1" X 2" X 1" X 18" CHANNEL IRON

1/4" X 1" X 20" BAR

1/2" X 7-1/2" BOLT

INVERTED HEAD FROM 30-GALLON HOT WATER TANK

1/2" X 1/2" X 11-1/16" BAR

3/8" PIPE TO 3/8" HOSE BARB

1/4" X 12" ROD BENT TO SHAPE

1/8" X 4" X 52-1/2" FLAT STEEL BAND

KAST-SET CAST-ABLE REFRACTORY

2" X 5" SLASH-CUT PIPE

3/8" X 6" ROD

2" PIPE CAP

1/2" X 6" BOLT, NUT, AND WASHERS

1" PIPE TEE W/1/2" CENTER SPLIT

MAKE ASBESTOS GASKETS

WELD BOLT HEAD TO 1/4" COUPLER

5/16" X 3" BOLTS W/DOUBLE NUTS

The gasifier is bolted in the bed of the pickup behind the cab, while the condenser and the filter ride outboard on one of the steps.

WOOD GASIFICATION

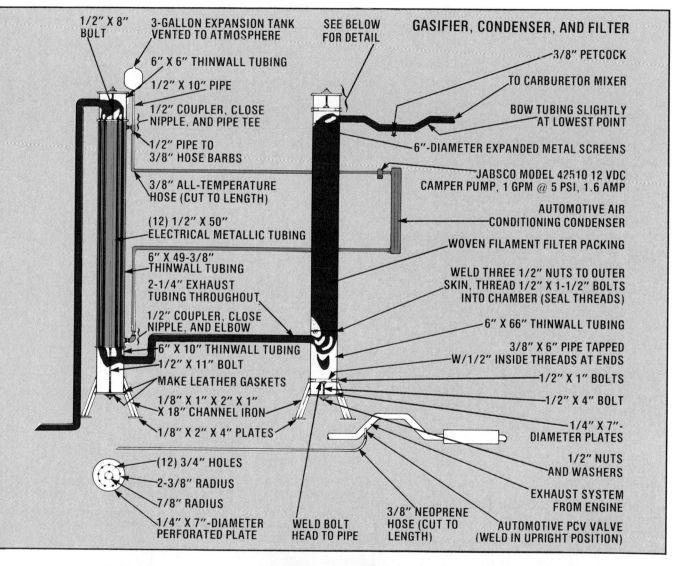

GASIFIER, CONDENSER, AND FILTER

- 1/2" X 8" BOLT
- 3-GALLON EXPANSION TANK VENTED TO ATMOSPHERE
- SEE BELOW FOR DETAIL
- 3/8" PETCOCK
- TO CARBURETOR MIXER
- 6" X 6" THINWALL TUBING
- 1/2" X 10" PIPE
- BOW TUBING SLIGHTLY AT LOWEST POINT
- 1/2" COUPLER, CLOSE NIPPLE, AND PIPE TEE
- 6"-DIAMETER EXPANDED METAL SCREENS
- 1/2" PIPE TO 3/8" HOSE BARBS
- JABSCO MODEL 42510 12 VDC CAMPER PUMP, 1 GPM @ 5 PSI, 1.6 AMP
- 3/8" ALL-TEMPERATURE HOSE (CUT TO LENGTH)
- AUTOMOTIVE AIR CONDITIONING CONDENSER
- (12) 1/2" X 50" ELECTRICAL METALLIC TUBING
- WOVEN FILAMENT FILTER PACKING
- 6" X 49-3/8" THINWALL TUBING
- WELD THREE 1/2" NUTS TO OUTER SKIN, THREAD 1/2" X 1-1/2" BOLTS INTO CHAMBER (SEAL THREADS)
- 2-1/4" EXHAUST TUBING THROUGHOUT
- 1/2" COUPLER, CLOSE NIPPLE, AND ELBOW
- 6" X 66" THINWALL TUBING
- 6" X 10" THINWALL TUBING
- 3/8" X 6" PIPE TAPPED W/1/2" INSIDE THREADS AT ENDS
- 1/2" X 11" BOLT
- MAKE LEATHER GASKETS
- 1/2" X 1" BOLTS
- 1/8" X 1" X 2" X 1" X 18" CHANNEL IRON
- 1/2" X 4" BOLT
- 1/8" X 2" X 4" PLATES
- 1/4" X 7"-DIAMETER PLATES
- (12) 3/4" HOLES
- 1/2" NUTS AND WASHERS
- 2-3/8" RADIUS
- 7/8" RADIUS
- EXHAUST SYSTEM FROM ENGINE
- 1/4" X 7"-DIAMETER PERFORATED PLATE
- WELD BOLT HEAD TO PIPE
- 3/8" NEOPRENE HOSE (CUT TO LENGTH)
- AUTOMOTIVE PCV VALVE (WELD IN UPRIGHT POSITION)

An expansion chamber allows for water level changes induced by heat. To assure trouble-free operation, the filter should be cleaned and the condenser drained periodically.

FUEL FROM THE FOREST

A Tree-Powered Truck! (continued)

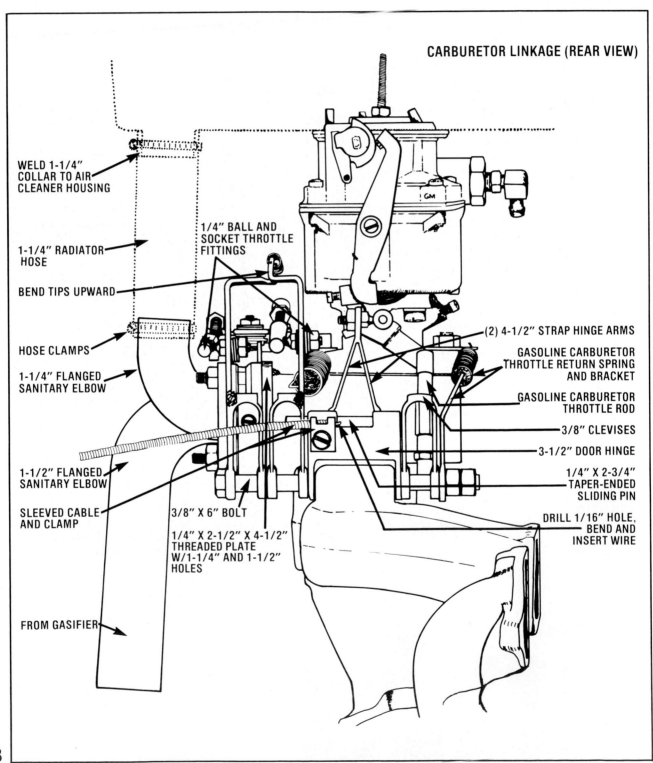

CARBURETOR LINKAGE (REAR VIEW)

WELD 1-1/4"
COLLAR TO AIR
CLEANER HOUSING

1-1/4" RADIATOR
HOSE

BEND TIPS UPWARD

HOSE CLAMPS

1-1/4" FLANGED
SANITARY ELBOW

1-1/2" FLANGED
SANITARY ELBOW

SLEEVED CABLE
AND CLAMP

FROM GASIFIER

1/4" BALL AND
SOCKET THROTTLE
FITTINGS

3/8" X 6" BOLT

1/4" X 2-1/2" X 4-1/2"
THREADED PLATE
W/1-1/4" AND 1-1/2"
HOLES

(2) 4-1/2" STRAP HINGE ARMS

GASOLINE CARBURETOR
THROTTLE RETURN SPRING
AND BRACKET

GASOLINE CARBURETOR
THROTTLE ROD

3/8" CLEVISES

3-1/2" DOOR HINGE

1/4" X 2-3/4"
TAPER-ENDED
SLIDING PIN

DRILL 1/16" HOLE,
BEND AND
INSERT WIRE

valve, and the far right U-clasp puts the gasoline carburetor's accelerator rod in motion.

The broad arm in the middle of the assembly functions as a master control and is connected to the truck's "gas" pedal. The component is merely half of a 3-1/2″ door hinge with two "feet" welded to its flat end (to allow it to pivot), and a two-sided yoke—made from a pair of 4-1/2″ strap hinges—welded to the opposite, curved edge. A sliding, taper-ended pin, measuring 1/4″ X 2-3/4″ and and controlled by a sleeved cable terminating at the dashboard, rests in the loops of the door hinge. As the pin moves laterally, this simple arrangement regulates the operation of either the gasoline or the wood-gas throttle (or both). In addition, a slotted bracket attached to each producer gas clevis allows both the fuel and the air "damper" travel within the carb throttle body to be adjusted, to achieve the air-to-gas combustion ratio.

To provide further flexibility, you can modify the linkage once more and make a great difference with respect to the engine's performance under the sorts of unpredictable traffic conditions that the everyday driver might encounter. Instead of allowing the sliding control pin to move both "smoke fuel" clevises at the same time, drill an access hole in only the right shoulder of the air mix arm and fasten a short spring between the lever's slotted bracket and that of its (wood gas) neighbor. Next, determine the position of both the air and wood-gas throttle rods when the "flap" valves within the two-port carb body are fully open, and then carefully install a two-stage spring on the air control lever so the secondary, or stouter, coil will come into play exactly at that point in the progression.

The final modifications include tack-welding a small stop onto the shaft-mounted arm of the producer gas throttle valve to prevent it from revolving past its full-open position . . . and driving a 1/8″ X 3/4″ roll pin into a hole in the upper surface of the carburetor body, which allows the air "damper" to travel beyond its maximum flow location, yet keeps it from going so far as to close it again *completely*.

In use, the dual-fuel setup is very effective. With the dashboard control cable pushed in, the truck runs solely on gasoline. But when the handle is pulled to its midway position, both the wood-gas and the petrol throttles function . . . allowing the motorist to pull away while rapidly bringing the gasification unit to a good fuel-producing temperature.

Once the coals are up to heat, the cable can be fully extended, and the engine can run on producer gas alone. Since in this configuration, there isn't sufficient draw to influence the carburetor, little or no gasoline enters the manifold through that atomizer's idle circuit. Furthermore, the adjustable progressive throttle assembly allows the engine to receive the proper air/wood gas ratio at all times, and the driver can feel the powerplant's response to any given traffic situation and can make corrective changes with the accelerator pedal.

As an alternative, you might prefer a simple tubular dual-fuel system (using either gasoline or processed smoke) rather than this more intricate setup. Though its performance isn't quite as good as that provided by the two-barrel carburetor body system, the unit is easier to build, and thus might be more attractive to some. This device appears to be ideal for a stationary, fixed-

A Tree-Powered Truck! (continued)

speed engine—such as that used on a generator—because the RPM level could be set, and then the air/fuel ratio adjusted as required *on site* for the best economy. And it may be possible to connect the control arms of the tubular carburetor's air and wood-gas valves in such a manner that both move simultaneously, thus providing an adequate flow of both substances during operation of the truck.

To facilitate the use of either gasoline *or* wood gas, fabricate a valved "T" from muffler tubing to fit the manifold inlet. The original carburetor is attached to one "arm" of this junction pipe, and the throttle body (containing just the butterfly and idle circuitry) from a scrap carb is fastened to the conduit's opposite end. A second piece of tubing links this valve housing with another just like it—to which an air cleaner has been installed

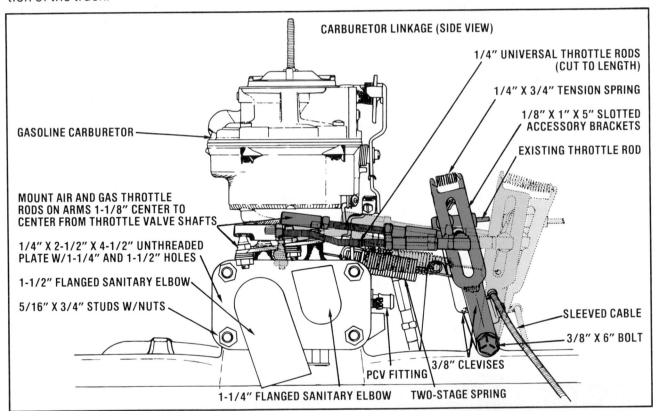

CARBURETOR LINKAGE (SIDE VIEW)

1/4" UNIVERSAL THROTTLE RODS (CUT TO LENGTH)

1/4" X 3/4" TENSION SPRING

1/8" X 1" X 5" SLOTTED ACCESSORY BRACKETS

EXISTING THROTTLE ROD

GASOLINE CARBURETOR

MOUNT AIR AND GAS THROTTLE RODS ON ARMS 1-1/8" CENTER TO CENTER FROM THROTTLE VALVE SHAFTS

1/4" X 2-1/2" X 4-1/2" UNTHREADED PLATE W/1-1/4" AND 1-1/2" HOLES

1-1/2" FLANGED SANITARY ELBOW

5/16" X 3/4" STUDS W/NUTS

SLEEVED CABLE

3/8" X 6" BOLT

3/8" CLEVISES

PCV FITTING

1-1/4" FLANGED SANITARY ELBOW

TWO-STAGE SPRING

Scraps of lumber make excellent fuel for a firewood-powered vehicle, and they are often available by the pile.

WOOD GASIFICATION

—and gas from the generator is introduced between the two "controlling" components. Finally, a manually operated sliding plunger is installed within the tubular "T" to serve as a fuel-mode selector.

Here's how the system works: If the engine is to be run on gasoline, the fuel selector plunger is pushed forward to block any "feed" from the gasifier and to let the petrol carburetor function normally. However, when the plunger is drawn to the rear, the gasoline system is neutralized, allowing the wood gas to take over. With this done, the air/fuel mixture can be controlled by adjusting the throttle "flap" beneath the gas system's air filter, and the vehicle's speed can be governed in a normal fashion via the second butterfly valve, which is connected to the accelerator linkage. By adapting this kind of setup, it's possible to maintain drivability in either fuel mode with a minimum of in-transit fuss . . . and to reap the additional benefits of [1] avoiding periodic maintenance on the gasoline carburetor due to wood-gas residue buildup, and [2] supplying the engine with necessary filtered air, regardless of which fuel the engine is using.

Like any device that depends upon "unfamiliar" technology, this scrap-fired pickup does take some getting used to . . . and admittedly, it has some drawbacks that should be carefully considered. For one thing, the power output of the engine is noticeably reduced. Even so, the truck starts easily, idles and runs smoothly, keeps up with traffic, and can maintain speeds beyond the legal limit.

Second, and extremely important, is the danger from carbon monoxide gas that may find its way out of the cooker or the supply pipes. If care is taken in the construction of the system and if no leaks are present, the engine will consume the toxins . . . exhausting CO emissions which register 33% lower than those given off in the gasoline mode. (Hydrocarbon emissions are reduced by half!) However, breathing the fumes from even a small breach in the fuel feed pipe, or the smoke given off when the gasifier is reloaded, could cause severe headaches and dizziness. And exposure to the odorless, colorless, and highly poisonous carbon monoxide fumes can bring about collapse or even death.

Finally, wood power simply isn't as convenient as are the more conventional methods used to get you down the road. The filter medium should be replaced after several hundred miles, and the condenser must be ramrodded with a shotgun cleaning brush (or flushed with a garden hose) and drained periodically.

It would be difficult to find fault with the price of the fuel, however. Virtually every bit of wood can be obtained for free . . . and you can burn everything from cast-off shop scrap to dead roadside brush to leftover building materials (often available by the pile!). And this thrifty vehicle, with a full wood supply and passenger load, goes about one mile on a pound of chunks, so the mileage figures out to some 75 miles per tankful.

Additionally, folks who might be concerned about the smoke's effect on the engine can take solace in the fact that valve seats and the combustion chamber stay surprisingly clean.

It would be difficult to imagine anyone not willing to investigate alternatives now, in preparation for the future. And, of the motor fuel options available today, wood gas is certainly one of the easiest and least expensive to produce and use.

A substantial supply of fuel can be loaded into the pickup's bed behind the gas generator.

HEAT PUMPS REALLY WORK

Many houses in North America seem to have been designed in such a way that it's impractical to retrofit them with solar-powered heaters, woodburning space heaters, or other low-cost methods of warming a dwelling. For these residences, central heating is often a "given", so it's only natural that more efficient and less costly methods of heating—and cooling—these homes have been sought.

The heat pump seems to be one such answer to lowering utility bills, and through recent research, these units—which extract heat from one medium and deposit it in another—have become quite efficient . . . and more and more popular.

Variations on the more basic air-to-air exchangers have also developed, using groundwater or other thermal sources as exchange mediums. Electrically driven heat pumps are combined with other means of heating, such as solar collectors and thermal mass, for even greater energy cost reductions than would be possible with a heat pump alone.

HEAT PUMPS REALLY WORK!

How the Heat Pump Works

Since 1973 when oil and gas prices took the first of several astronomical leaps, the heat pump has become increasingly popular as a home heating and cooling device. Functioning on the same principles of refrigeration as does the modern icebox, these dual function appliances cost, on the average, about 20% less to operate than do gas, oil, and electric furnaces, and in some parts of the country, the savings can reach as high as 40%.

Unlike conventional heating systems, which generate heat from coal, oil, gas, wood, or electricity, a heat pump itself generates no heat but collects warmth from one area and transfers, or pumps, it to another. In its heating cycle, a heat pump extracts warmth from the outside air or from some other medium such as water and transfers those Btu into a dwelling. In the cooling cycle, the unit takes heat from the inside air and deposits those unwanted Btu outside.

The operation of a heat pump depends on a basic principle of thermodynamics under which a substance changing from a liquid to a gaseous state absorbs heat while a substance changing from a vapor to a liquid gives off heat.

Basically, a heat pump consists of an evaporator coil, a condenser coil, a compressor, a reversing valve that changes the roles of the two coils, and restricting valves that limit the flow of the refrigerant with which the system is filled. Fans move air over the coils, where that air either loses or gains heat, though in a water-to-air unit one of these would be replaced by a pump.

In air-to-air heat pumps, the refrigerant is continuously changed from a gas to a liquid and back again as it is circulated through the system by the compressor. Refrigerant entering the evaporator coil expands and becomes a vapor, absorbing warmth from air passed over the coil by a fan. The vapor is then moved by the compressor to the condenser coil where it changes to a liquid state and releases the heat it gained in the evaporator coil. This heat is carried away from the coil by a fan. From the condenser coil, the refrigerant travels through a restricting valve into the evaporator coil where it vaporizes, completing the cycle.

Whether a heat pump is functioning as a heater or as an air conditioner depends on the reversing valve. This mechanism controls which direction the refrigerant flows through the coils and determines which coil acts as the evaporator and which as the condenser.

An air-to-air heat pump will have one coil located outside the building or otherwise exposed to outside air, and the other will be located inside the dwelling where air blown over it can be distributed by ductwork throughout the residence.

When the unit is functioning as a heater, the outside coil serves as the evaporator, where the refrigerant picks up heat from the air. The inside coil acts as the condenser, where the vapor is liquefied and gives up the heat it gained outside.

In the cooling mode, the roles are reversed, and the inside coil is the evaporator, or heat-absorbing end of the system, while the unwanted warmth is deposited outside as the coil there acts as the condenser.

The amount of power needed by a heat pump to extract heat from the air depends on the temperature of the air from which the unit must draw that warmth. For example, it takes a great deal more work to extract a given number of Btu from 10°F air than it does to remove the same amount of heat from 30°F air. Thus, a heat pump is obviously more practical for heating homes in the

GETTING WARM FROM COLD

milder parts of the country than it is for those in the far northern areas.

In an effort to increase the practicality and popularity of heat pumps and to decrease their overall cost per Btu, developments have been made that include using well water or water from a solar heated pool as a heat exchange medium for the "outside" coil (which can be located in the building as part of a single unit). There are also closed loop systems in which water, some other liquid, or even air is circulated around the "outside" coil and through a long loop of pipe or tubing buried in the ground or anchored to the bot-tom of a lake (or some other body of water large enough to maintain a fairly stable temperature). In fact, almost any thermally stable mass of adequate size can be used as a heat transfer medium.

The cost-effectiveness of a heat pump can be determined only on an individual basis and depends on several factors, including the location where it might be used, the price of other fuels available in the area, and even the relative cost of the unit itself. Whether or not a heat pump is the "right" system for a given home can be ascertained only by comparing the unit's long-term cost per Btu with that of the alternatives.

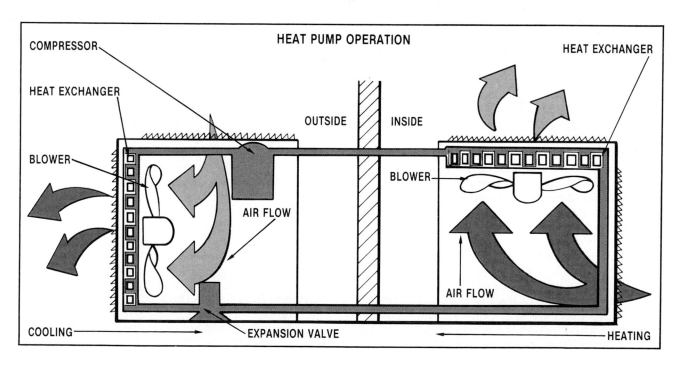

HEAT PUMP OPERATION

HEAT PUMPS REALLY WORK!

A Well Water "Furnace"

Few people would blink an eye if they heard about a system designed to utilize cool water to fight the summer heat, but those same individuals might just think it a little strange to propose using that same cool water to heat a home during the colder months. Nonetheless, that's precisely what happens in a water-to-air heat pump.

These units are designed so that during the heating season, they can extract heat from well water and transfer that warmth to the living area of a dwelling, and so that during the time of year when air conditioning is desired, they can extract heat from the air inside the house and deposit it in the water pumped from the earth.

Groundwater remains at a fairly constant temperature throughout the year, averaging about 56°F in the United States, and there's plenty of warmth that can be drawn from water in the 50° to 60°F range. In fact, a heat pump can extract warmth from water right down to the liquid's freezing point of 32°F when it becomes ice.

The advantage of a water-to-air heat pump over an air-to-air unit is that during most of the year, the liquid-using type does not have to work as hard to transfer the same amount of heat. In the dead of winter, it requires less energy to extract a given number of Btu from 56°F water than it does to wrest the same quantity of heat from less-than-56°F air. In the summer, it's easier to deposit a given number of Btu into 56°F water than to dump the same amount of warmth into greater-than-56°F air.

In one specific water-to-air exchanger installed in a 1,700 square foot home, 60°F well water is pulled into the unit at a rate of eight gallons a minute. The well itself produces about 50 gallons of water a minute, so there is no chance of the

The working "heart" of the water-to-air heat pump is a compact unit located in the dwelling.

heat pump's drawing the well dry, especially since the unit operates intermittently under the control of the home's thermostat. After the warmth has been extracted, the water—whose temperature has dropped to about 45°F—is discharged into a drain field (some areas require that the liquid be returned to the groundwater supply by means of a second well). Meanwhile, the heat drawn from the water is blown into the living area of the house.

In a comparison between two similar houses, the power bills for one heated with a water-to-air unit averaged about 20% less than those of the other dwelling, which was warmed and cooled by an air-to-air heat pump.

The well cap in the yard is low to the ground and is unobtrusive. Groundwater is piped to the heat pump through pipes beneath the lawn.

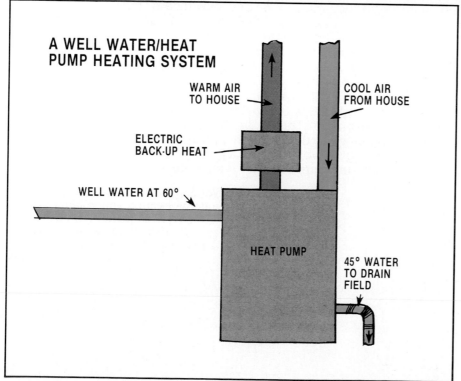

A WELL WATER/HEAT PUMP HEATING SYSTEM

WARM AIR TO HOUSE

COOL AIR FROM HOUSE

ELECTRIC BACK-UP HEAT

WELL WATER AT 60°

HEAT PUMP

45° WATER TO DRAIN FIELD

Installed in a conventional house, the water-to-air heat pump cost about 20% less to operate than did an air-to-air unit in a similar house nearby. The diagram outlines the heat pump's operation during the cooling cycle.

HEAT PUMPS REALLY WORK

The Heater That You Swim In

During the winter, water from the swimming pool is circulated through the roof-mounted solar collector where it picks up heat that is later transferred to the dwelling's living area by two heat pumps.

What do heat pumps, a swimming pool, and rooftop solar collectors have in common? Well, in a hilltop house near Asheville, North Carolina, they have teamed up to provide warmth in winter and cooling in summer for the energy-efficient dwelling's 3,000-square-foot living area.

In this comprehensive system, water from the pool—which is covered with an insulating cap during the heating season—is pumped through solar collectors where it is warmed by the sun. The heat pumps, one of which serves each of the house's two stories, then extract warmth from the water and transfer that heat throughout the dwelling.

A pool-based solar heating system will work most efficiently if it functions in tandem with a home designed for solar heat, one that is extremely well insulated. And almost all of the dwelling's windows (double glazed) and other glassed surfaces (duo-pane sliding doors) should face south, thus acting as additional "free" passive thermal energy collectors during brisk winter days. The house, in short, should be planned with an eye toward snug cold-weather comfort with only a minimal need for backup heat. A home similar in size to the one shown here could easily be heated with a 30,000-gallon pool and a few conventional flat-plate solar collectors.

Though very few flat-plate collectors are needed in the setup, it's necessary to use the absolute best: five 5′ X 16′ units glazed with tempered glass. (In the home here, a sixth collector that measures 5′ X 8′ provides household hot water.)

Thanks to the extreme efficiency of the heat pumps which deliver warmth from the heat reservoir (that 30,000-gallon pool) to the house, the required water temperature can be much lower than

Working together to form a complete home heating and cooling system, are a 30,000-gallon swimming pool (which can also be used for recreation), solar collectors, and a pair of heat pumps.

GETTING WARM FROM COLD

HEAT PUMPS REALLY WORK

The Heater That You Swim In (continued)

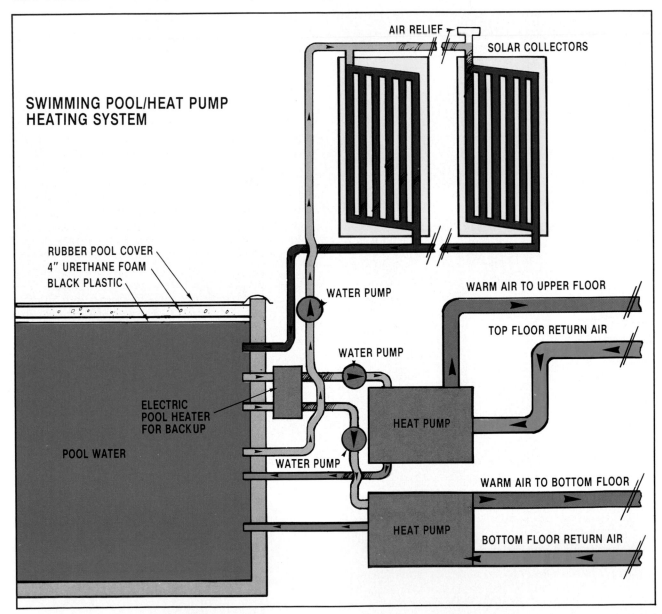

SWIMMING POOL/HEAT PUMP
HEATING SYSTEM

AIR RELIEF

SOLAR COLLECTORS

RUBBER POOL COVER
4" URETHANE FOAM
BLACK PLASTIC

WATER PUMP

WARM AIR TO UPPER FLOOR

TOP FLOOR RETURN AIR

WATER PUMP

ELECTRIC
POOL HEATER
FOR BACKUP

HEAT PUMP

POOL WATER

WATER PUMP

WARM AIR TO BOTTOM FLOOR

HEAT PUMP

BOTTOM FLOOR RETURN AIR

GETTING WARM FROM COLD

that usually called for in a solar heating system's storage unit. It's better, in this case, to circulate a lot of water through a flat-plate collector quickly, and to heat it all just a little bit, than to pump a mere trickle of water through the same collector and to heat the smaller volume of the fluid intensely.

Also, it's possible for a pool that is set in the ground to draw some "free" heat directly from the earth.

But the real heart of this solar heating system is the pair of water-to-air heat pumps that extract warmth from the pool, which during cold months should be covered with a sheet of plastic, then four inches of urethane foam insulation, and topped with a rubberized cover. These units concentrate the extracted warmth, use it to heat air, and then blow that warm air into the house.

The two such pumps (each rated at four tons of capacity) are able to keep the 3,000-square-foot home comfortably warm—and should be able to

A wooden deck and balcony on the south side of this house afford the residents a panoramic view of the southern mountains.

do so in any climate—even if the temperature of the water in the covered pool drops to 50°F or lower. Water at that temperature still contains a considerable amount of heat, and the pumps, if necessary, can extract much of that warmth.

The heat exchangers would have to work hard to do their job if the storage unit's temperature fell much below 50°F. However, even in the most bleak, windswept winter, a well-insulated pool's temperature will rarely—if ever—dip below 50°F.

Controlling the system is straightforward. There are no supersophisticated computers to go haywire, and there is no expensive antifreeze or heat transfer fluid to leak into the pool and contaminate its water. Ordinary thermostats control every circuit in the system. The same water that's stored in the pool and pumped through the rooftop collectors flows through the heat pumps, which transfer warmth to the house as desired. The water is automatically dumped from the collectors and into the pool each night during the winter to keep it from freezing.

In the summer, this heating system changes roles and becomes a cooling system, but there's an obvious bonus. The storage tank is stripped of insulation and is transformed into what it was originally: a refreshing swimming pool.

If the house becomes too hot, it can be airconditioned by running those heat pumps backward to extract heat from the house and put it into the pool. And if that makes the water in the pool too warm, it can be circulated up through the rooftop collectors at night to let the excess heat drain away into the cool nocturnal air.

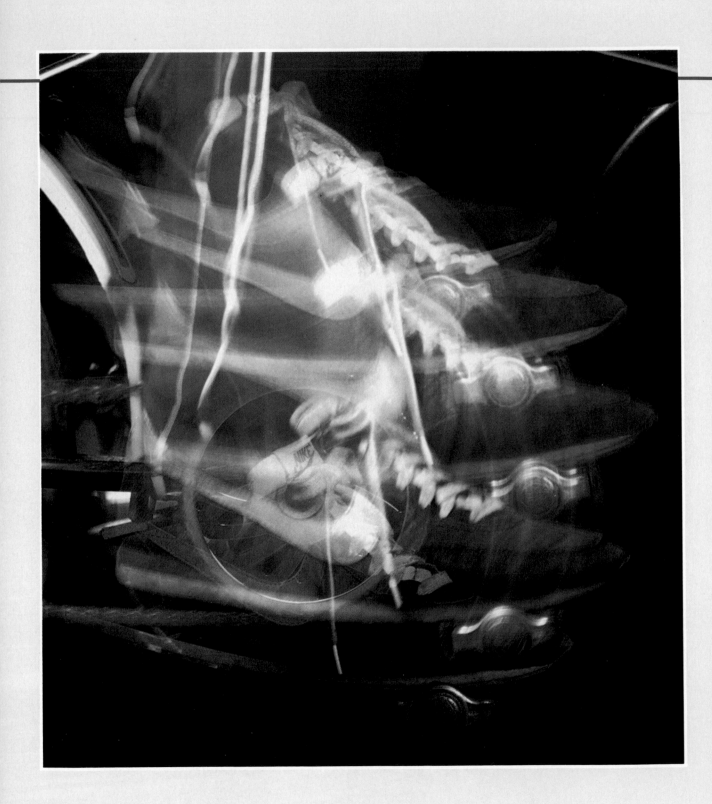

POTPOURRI

So far we've explored the largely untapped potential of the sun, wind, and water, and of methane, fuel alcohol, heat pumps, and wood . . . and discussed the need to *store* home-generated electricity in banks of batteries. Well, if all of the simple-but-ingenious ideas presented in the preceding chapters haven't impressed you with the vast range of alternatives open to us all, the material that follows most certainly will.

This final chapter spans a range of subjects and of complexities. The ideas presented include tips on evaluating specific energy-saving alternatives for your own location (so you can predict savings *before* laying down cold hard cash) . . . a simple means of turning your fireplace into a home cooling system . . . a design for an anyone-can-do-it setback thermostat system . . . the instructions for a somewhat complicated—but darned efficient—carburetor conversion that'll let a "normal" car run well on alcohol fuel . . . and more. So dip into our final grab bag of ideas . . . if you don't find anything here that you can use, it's a good bet that you'll at least find a few sparks to ignite ideas of your own!

AN ENERGY POTPOURRI

A Carburetor Cocktail

Introducing an alcohol or alcohol/water mist into the intake of a vehicle's engine can result in increased mileage, improved acceleration, and reduced preignition. In fact, depending on the type of vehicle, the operator's driving habits, and the method of introducing the fluid, either mileage gains of about 15% or notable improvements in performance may be possible on an otherwise stock engine.

There are a number of different approaches to controlling the volume and timing of misting, but the two most common are manifold vacuum and smog-pump compression. In the interest of serving those people with cars that predate the smog-pump era (though the approach described is perfectly applicable to later models, too), the following information will help in building a simple vacuum-controlled mister. Though the vacuum-controlled system will not provide the ideal metering of which the pressurized setup is capable, it will introduce a sufficient amount of liquid into the gas-air mixture to reduce overall combustion temperature peaks while smoothing and boosting the pressure rise in the engine's combustion chambers.

The system shown in the accompanying photos has been installed on two vehicles: a 1969 Volkswagen Type I and a 1979 Subaru Brat. Before the mister was added to the VW beetle, the car's engine was subject to occasional fits of preignition. The introduction of either alcohol or alcohol/water vapors in a 1-to-4 ratio into the manifold reduced the "pinging" significantly and improved acceleration noticeably. Unfortunately, there was no mileage improvement.

On the Brat, however, there was a measurable increase in fuel economy. Before installation of the water mist setup, the minipickup logged 25.8 miles per gallon. And after conversion, it registered it 29.4 MPG. This computes roughly to a 14% improvement. Oddly enough, there was no detectable difference in acceleration. (Knocking was never a problem in the Subaru.)

While it's difficult to draw any solid conclusions from this, one possible explanation for the different results is that a mister may work better on a water-cooled engine (such as the Subaru's) than it does on an air-cooled powerplant (such as the Volkswagen's). A specific explanation could include the possibility that the VW's intake manifold, which is exposed to the air and preheated by an exhaust gas riser, may be cooled too much by the use of a misting device.

To build a mister like this one, you'll need a lidded jar of between one pint and one quart capacity, a Fram CG-13 fuel filter (to serve as the bubbler), a foot of 1/8″ soft copper tubing, a brass aquarium valve, three to five feet of appropriately sized neoprene tubing, some 50/50 solder, and a propane torch.

Drill or punch two 1/4″ holes about an inch from the edge of the jar's lid and opposed to each other. Then trim off an 8″ and a 2″ length of copper tubing. Solder the fuel filter to the end of the 8″-long pipe, slip the tube through one of the holes in the cap, and position it so that the filter will be within 1/4 inch of the jar's bottom when the lid is tightened. Insert the shorter piece of pipe through the other hole, so that it protrudes only 1/2 inch into the jar, and solder the two tubes to the jar lid.

Secure the aquarium valve right to the edge of the jar's lid, or—if it's more convenient—bolt or solder it to a panel in the engine compartment.

Whichever approach you choose, try to locate the jar close to both the carburetor and the distributor to reduce the amount of tubing needed for connecting the system's components.

Now, plumb the three-way aquarium valve into the vacuum line between the carburetor and the distributor. Then connect the short tube on the bubbler bottle to the remaining hose barb on the valve. Be certain that all the lines fit very snugly: Even a small leak can adversely affect the performance of both the mister and the mechanism to which the vacuum line is connected.

Fill the jar to about one inch below the inlet tube with the solution you've chosen, turn the valve all the way off, and start the car's engine. Then slowly open the valve until a very gentle bubbling occurs at idle. Rev the motor a few times to be certain that the bubbling doesn't become so violent that liquid is drawn directly into the tube, adjust the valve accordingly, and you're ready to go.

It may take a couple of tanks of gas to realize the mister's full potential for fuel economy, and during that time you'll probably be able to make fine adjustments to the valve to improve the system's performance. But from then on, it should just be a matter of refilling the jar each time you stop at the gasoline station.

This homemade alcohol mister, positioned between the carburetor and the distributor, will improve the performance of your car's engine.

ENERGY POTPOURRI

Pedal Power

Without question, the most energy-efficient mode of transportation available today is the bicycle, on which people can move from one place to another nearly five times as efficiently as they can by walking. And because of this spectacular efficiency, many tinkerers have put bicycles to use doing work other than getting from here to there. The antecedents of such a practice include the treadle sewing machines and the pedal-powered lathes and sharpening wheels of our forebears. Other areas of usefulness still exist for the old two-wheeler: Pedal power can pump water and generate electricity.

Quality ten-speed bicycles are carefully designed to make the best use of the power people can supply with their legs. If one wants the most efficient pedal-energy device possible, he or she should start with a well-built cycle. This doesn't mean spending several hundred dollars to buy a brand new lightweight racer. Usually the parts for a foot-cranked generator can be found in junkyards, or you may possibly find a discarded 10-speed in the scrap pile at a hardware store.

Once you acquire the gear mechanisms, the construction of such a powerplant is mostly a matter of providing the "cycle's" chassis with a stable platform. Start by detaching the wheels from both ends of the bike, and then loosening and removing the rim and spokes from the rear wheel hub. To accomplish this task, you'll need to remove the assembly from the frame, and either pull the tire and tube from the wheel rim and unscrew each spoke (use a suitable spoke wrench or a screwdriver modified to slip into the spoke nipple), or cut each spoke and remove it from the hub. The next step is to reinstall the rear hub and axle.

At this point you should remove the bicycle's front fork assembly. You can attach a metal pipe base directly to the fork legs, but the wheel supports of most bicycles are designed to flex slightly to absorb road shock and will provide a rather wobbly mount for your cycle-power unit. Instead, a 31″ length of 1″ E.M.T. (electrical metallic tubing) is welded to the conduit legs, run through the cycle's steering head, and eventually—once the assembly has been leveled—bolted securely in place by means of a set screw.

The front legs of the frame are formed from a 29″ length of 3/4″ E.M.T. with a six inch radius bent 90° at each end. The same curves should be made on the base's rear legs, but a 76″ piece of E.M.T. is used for this. Then center the bends on this piece 28-1/4 inches from each end of the tubing. Either a conduit bender or an adjustable arc roller is needed to shape both the front and the rear legs.

When the leg assemblies are ready, cut two 49″ lengths of 1/2″ E.M.T. to serve as connectors between the front and rear sections. Then drill 1/4″ holes at points an inch from each end of both conduits, making sure that the bores are parallel, and bolt the frame loosely together with 1/4″ X 2″ bolts, washers, and nuts. (After establishing just how the cycle will fit into the base, you can snug the four bolts down.)

Now lift the stripped bicycle chassis into position and determine the most convenient way to secure it to the conduit framework. In the model here, the bracket that had been used to mount the rear brake lined up very well with the top of the back legs, so the two assemblies could be easily bolted together. (Bear in mind that there is a good deal of margin to work with. The front of

By combining a bicycle, a battery, and an automobile alternator, you can pedal-power small appliances in your home.

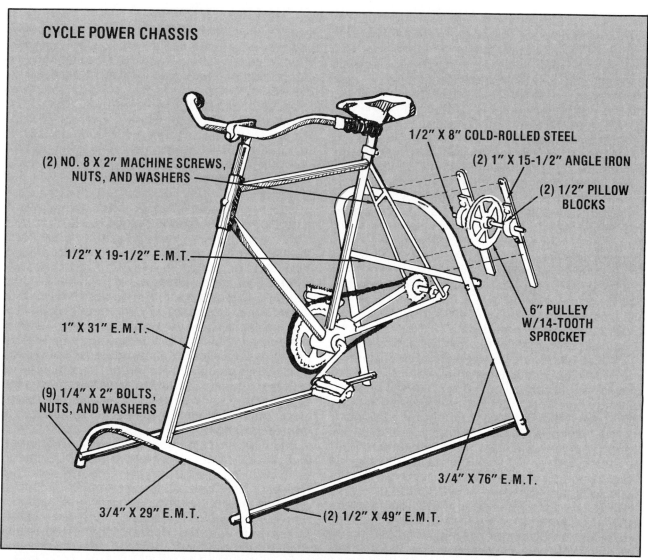

CYCLE POWER CHASSIS

(2) NO. 8 X 2" MACHINE SCREWS, NUTS, AND WASHERS

1/2" X 19-1/2" E.M.T.

1" X 31" E.M.T.

(9) 1/4" X 2" BOLTS, NUTS, AND WASHERS

3/4" X 29" E.M.T.

(2) 1/2" X 49" E.M.T.

1/2" X 8" COLD-ROLLED STEEL

(2) 1" X 15-1/2" ANGLE IRON

(2) 1/2" PILLOW BLOCKS

6" PULLEY W/14-TOOTH SPROCKET

3/4" X 76" E.M.T.

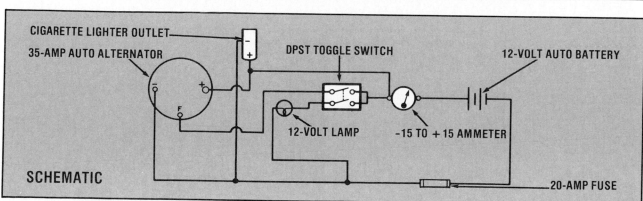

CIGARETTE LIGHTER OUTLET

35-AMP AUTO ALTERNATOR

DPST TOGGLE SWITCH

12-VOLT AUTO BATTERY

12-VOLT LAMP

−15 TO +15 AMMETER

20-AMP FUSE

SCHEMATIC

247

ENERGY POTPOURRI

Pedal Power (continued)

the bike can be raised or lowered—on the 1″ E.M.T.—for leveling purposes.) When you're satisfied with the fit, drill the appropriate holes, bolt the bicycle into place, and cinch up the 1/4″ bolts on the E.M.T. framework.

The chassis should now be relatively steady, but to add a little extra rigidity, run a 19-1/2″ piece of 1/2″ E.M.T. across the rear legs, 19-1/2 inches above the floor. This tube bolts both to the legs and to the rear fork of the bicycle and helps keep the device from rocking forward.

Because the bicycle chain system is not suited for directly driving most machinery, a shaft is mounted above the rear axle to a V-belt and pulley arrangement. Two 15-1/2″-long pieces of 1″ angle iron are used to mount the pillow blocks that support the 1/2″ pulley shaft. The bottoms of the angle iron sections bolt directly to the bicycle's rear axle, and the tops are attached to the horizontal section of the rear legs. The pulley itself is a 6″ type A (the largest that will fit into the space available) and is driven by a 14-tooth sprocket compatible with the bicycle's chain.

Though the various projects that someone may want to undertake using this cycle-power system will require different rotational speeds to operate correctly, the vast majority will need more speed than a person will be able to develop at the crank. Therefore, mount the cycle's primary chain permanently on the largest front and smallest rear sprockets. The ratio achieved by this, combined with the additional step-up supplied by going from the largest rear sprocket to the 14-tooth spur on the pulley shaft, will increase the speed by a factor of about six.

The remaining components needed for a complete cycle-power system are a willing pedaler and an appliance to operate. While there are a number of different implements that can be powered with a pedaled setup, this one was first used to pump water—for heat storage—to a remote solar greenhouse. At the rotational speeds generated with this bike, a piston-type pump was found to be the best choice. Though new piston pumps tend to be relatively expensive, it's usually possible to find a serviceable used unit. Once the pressed-back-into-service pump was primed, an enthusiastic operator could enable it to lift water as high as 25 feet and to deliver around 120 gallons per hour.

Beyond this specialized use for the old pedal-pusher, you can operate any 12-volt appliance with the bicycle by combining the cycle-power chassis and an automotive alternator to then generate electricity.

First, a flywheel must be added to the existing setup to help stabilize the pulses that are produced by the thrusts of the rider's legs. For several reasons, a junked Volkswagen flywheel is a good choice: It has a flat surface next to the ring gear upon which a belt can ride, and also, the unit's one-piece construction (which includes a nondetachable ring gear) and recessed face make the VW bug part nearly impossible to repair, so these used flywheels are available for next to nothing.

To install the steel disk, simply center and weld it onto your bike's rear hub. Then, in order to provide clearance for the flywheel, relocate the rear frame brace. The upper 19-1/2″ conduit crosspiece should be moved up five inches to a point 24-1/2 inches above the floor. Then add two braces between the axle ends and the bolts that connect the horizontal and vertical frame mem-

---

LIST OF MATERIALS

Basic Chassis

(1)	Salvaged 10-speed bicycle
(1)	1/2″ X 10′-long electrical metallic tubing (E.M.T.)
(1)	3/4″ X 10′-long E.M.T.
(1)	1″ X 3′-long E.M.T.
(1)	6″ type-A pulley and 14-tooth sprocket

(1)	V-belt
(2)	1/2″ pillow blocks
(1)	1/2″ X 8″ cold-rolled steel
(1)	1″ X 31″ angle iron
(9)	1/4 X 2″ bolts, nuts, and lock washers
(4)	1/4″ 1″ bolts, nuts, and lock washers
(2)	No. 8 X 2″ screws, nuts, and washers

bers. The combination will increase the rigidity of the chassis.

The roughly 12"-diameter flywheel allows for greater ratio increase than did the old chain and 6" pulley system. After a number of experiments were done with different drive pulley sizes on the alternator of a trial unit, this new flywheel, teamed with the power-generating device's stock 3" pulley, seemed to provide an almost ideal ratio.

Attaching an automotive alternator to your cycle-power frame involves only minor modifications to the basic bicycle unit. A 6" piece of 1" angle iron is bolted to the angle iron braces on the chassis, and brackets are then fashioned to allow the generator to pivot to provide belt tensioning capability. Then attach an 18"-piece of threaded rod to one of the alternator's mounting bosses, and slip the other end through a hole in a steel tab bolted to the column that supports the bike's seat. Threading two nuts up or down the rod adjusts the belt's tightness.

For wiring the system, just follow the accompanying schematic. The essential components amount to nothing more than a positive connection between a battery and the alternator, a wire connecting the negative terminals, and a field wire that can be switched on or off.

Though the basis of the circuit is quite simple, there are a couple of refinements which—while not absolutely necessary—are worthwhile. First, wire a very low-wattage bulb into the field circuit to indicate when field current is flowing. Leaving the switch on when the charger's not in use will discharge the battery through the field windings. To prevent the bulb from being blown by surge when the circuit is shut off, the fixture is wired parallel to the field through a dual pole, single throw (DPST) switch. Another option is an automotive ammeter, the readings of which can be either an inspiration or a discouragement, depending on one's enthusiasm for pedaling.

"Well," you're probably wondering, "can I generate all my household's electricity with a bicycle?" The answer is yes, if you have family members who are in excellent physical condition and if you don't use much electricity. The maximum that an experienced cyclist will produce is about 16 amps at 13.5 volts or about 200 watts. However, he or she can maintain the pace needed to produce that output for only a few minutes. A practical figure for continuous pedaling is 5 amps at 12.5 volts, or about 60 watts.

Parents who wish to limit their children's television time by making the youngsters produce the electricity for the tube should know that the unit will require roughly a minute's pedaling for every two minutes of watching a small 25-watt, 12-volt TV. The electricity must be generated beforehand and stored in the battery. Erratic power surges will disturb the picture and damage electronic components.

A wide variety of 12-volt appliances can be served by the bicycle generator and battery setup. And the storage capacity of the cells will actually allow you to operate equipment—for limited periods—that draws more than you're able to produce on a continuous basis. But remember, every watt that's removed from the battery will eventually have to be replaced, and the use of a power-hungry 1,200-watt tool or appliance for one hour will necessitate at least 20 total hours of pedaling to restore the battery's charge.

A bicycle generator can give a practical, though limited, boost to energy-saving households.

LIST OF MATERIALS (continued)

Addition of Alternator

(1)	Volkswagen type-1 flywheel		(1)	DPST toggle switch
(1)	1" X 1" X 12" angle iron		(1)	Cigarette lighter outlet
(1)	1/4" X 18" all-thread rod, and (4) nuts		(1)	12-volt, 50-mA lamp
(1)	3/8" X 5" bolt, nut, and washers		(1)	-15 to +15 ammeter
(2)	1/4" X 2" bolts, nuts, and washers		(1)	12-volt auto battery
(1)	35-amp (or smaller) auto alternator junction box		(1)	20-amp fuse
			(1)	20-amp fuse holder
				30' of hookup wire

AN ENERGY POTPOURRI

Let Your Fireplace Cool Your Home

Heat, however it's produced, can be a real nuisance at times. In most sections of North America, the intense heat of a summer's day can result in a sweltering bedroom at night, even though the temperature outdoors drops after the sun goes down.

Of course, you can alleviate the sleep-robbing discomfort with an attic fan that moves stagnant air out of the house and at the same time, draws night-tempered outside air indoors. The installation of such an appliance can be expensive, though, and will leave a poorly insulated spot in the ceiling where the opening is made.

However, there's a way to have the advantages of an attic fan without much expense or any loss of weatherproofing. Just purchase an inexpensive square "window" fan. After opening the damper in your chimney, position the fan within the aperture of the fireplace so that air can be pulled from the room and blown up the chimney. Then close off the opening remaining around the fan using scraps of cardboard, particle board, plastic foam, or other flat material. To really go first class, add a high, three-sided, hinged wooden baffle to limit the fan's draw to the especially warm air that collects near the ceiling.

Once the equipment is assembled, simply switch on the fan and enjoy the benefits of a costly attic exhaust without the attendant expense. It's amazing how much brighter summer days will seem when they follow nights of cool and restful sleep.

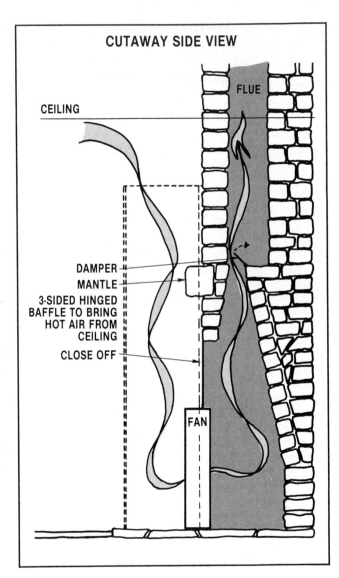

CUTAWAY SIDE VIEW

FLUE

CEILING

DAMPER
MANTLE
3-SIDED HINGED
BAFFLE TO BRING
HOT AIR FROM
CEILING
CLOSE OFF

FAN

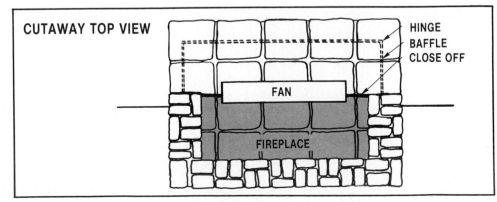

CUTAWAY TOP VIEW

HINGE
BAFFLE
CLOSE OFF

FAN

FIREPLACE

A Supersaver Stovepipe

Sometimes, it seems, winters get so harsh that icicles form on the inside of the windows just to get out of the cold! In such chilling situations, you'll do well to squeeze as much heat as you can out of the valuable hardwood logs that rapidly disappear into the gullet of your woodstove.

To extend the heat-radiating surface of your stovepipe, try adding the squared circle shown in the accompanying sketch. Standard eight-inch stovepipe—in the form of two T's, four 90° elbows, and a pair of straight sections—is easily connected to form the loop. The damper is located above the assembly, and from there, the pipe exits through the ceiling by way of an insulated triple-wall thimble. All the joints are fastened with sheet metal screws for security.

And, for safety's sake, remember that any device that salvages heat from a stovepipe is going to encourage the formation of creosote, and that could lead to a chimney fire. To prevent such an occurrence, clean the inside of the assembly whenever the sooty buildup gets to be more than 1/4″ thick. In addition, it's a good idea to replace all the pipe every second year. The latter may be an expensive practice, but good accident prevention is the best insurance.

AN ENERGY POTPOURRI

A Home Setback Thermostat

When long and frigid winter evenings roll around, many sleepers enjoy the economy and comfort of clicking their thermostats back to 60°F (or lower) and snuggling beneath an extra blanket or quilt. The cool night air against one's face, contrasted with the warmth provided by the additional bedclothes, seems to bring about an especially pleasurable sleep.

But oh, those first few dripping-wet steps out of the next morning's shower and into 60°F air can be a rude shock. Until now the only way around the AM goose bumps—short of an energy-eating space heater—has been either to rise at 4:00 AM and kick the thermostat back up, or to buy a costly electronic setback control. But here is a simple, and less expensive, way to achieve the same results as are possible with the commercially purchased setbacks.

Instead of controlling one thermostat with a timer, this unit uses two thermostats and a timer that controls an electromagnetic switch called a relay. When the relay gets power, by way of the 115V lines from the plug-in timer, it switches the 24V thermostat current from the furnace to the auxiliary thermostat. But when the timer kicks back off (in the morning), the temperature control reverts to the main thermostat. There are eight posts on a DPDT (dual pole dual throw) relay: two for the 115V household line from the timer (Nos. 2 and 7), two for the 24V wires from the furnace (Nos. 1 and 8), and a pair for each thermostat (Nos. 3 and 6 and Nos. 4 and 5). The numbers on the leads of the relay in the diagram are for a specific relay used to build a prototype, and anyone building this device should be aware that the configuration of the relay actually involved in construction may very well differ.

And for those occasions when you want to stay up a little later than usual and thus want the house to stay warm for an hour or so more, a manual override switch is helpful. Your timer may already be equipped with such a feature, but if not, just wire a toggle switch into one of the 115V lines. (If a pilot light is included in this circuit loop, it will indicate at a glance which thermostat is in operation.) Just be sure that the auxiliary has a lead rating similar to that of the main control.

While the setback control featured here is designed for heating units whose control mechanism is separate from household current, it can be easily adapted to home warmers such as baseboard heaters whose thermostats are wired directly into the 115-volt circuit. To make this adaptation, one leg of the 115-volt line feeding the heater is run first to one of the relay connectors where the line to the furnace is connected in the diagram and then from the other furnace lug to the heater. Also, the 24V heating thermostats must be replaced 115V in-line units.

As you can see, there's next to nothing involved in building this thermostat control. A couple of off-the-shelf parts and a little wire and solder are all that go into it. In fact, the only thing that takes less effort than building this setback thermostat is using it.

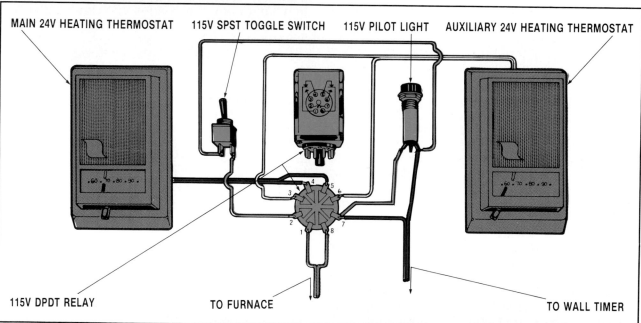

MAIN 24V HEATING THERMOSTAT — 115V SPST TOGGLE SWITCH — 115V PILOT LIGHT — AUXILIARY 24V HEATING THERMOSTAT

115V DPDT RELAY — TO FURNACE — TO WALL TIMER

Almost any discarded washing machine has a suitable 115V relay in it waiting to be scrounged—and building recyclers often have a supply of used heating thermostats they'd be glad to part with for a couple of dollars apiece—so by using some secondhand parts, you can cut the cost of constructing this unit.

LIST OF MATERIALS

1 heating thermostat (24V)
1 small appliance timer
1 DPDT 115V relay
1 pilot light (115V)
1 utility box
Wire, solder, and a household plug
1 SPST 115V toggle switch (optional)

AN ENERGY POTPOURRI

Renters Can Save Energy Too!

You say you live in a drafty, heat-leaking, energy-wasting budget-buster of an apartment or rental house? And the landlord or landlady is not inclined to invest in making energy-saving improvements to the building? Take heart, because there are several easy, low-cost ways in which you can reduce your dwelling's demand on power and increase your comfort.

Since 15% of all the power you use in your home goes for hot water production, you can begin your conservation efforts by making a beeline for that water heater. When you get there, first turn the appliance's 140°F (or higher) temperature setting down: 120°F is more than adequate for washing clothes, dishes, and people.

Second, wrap the tank in a blanket of 3-1/2" or 4" fiberglass insulation. (Be very careful to leave vents or flues at the top of gas- or oil-fired heaters unobstructed and to leave clearance around any relief valves, thermostat units, and burner or coil areas.) Tape the batts together or, for an installation that's easy to take with you when you move and less likely to cause compaction of the fibrous material, build a support structure from wire coat hangers and fit your insulation layers into these homemade brackets.

You can also save energy by outfoxing your furnace. Get a 24-hour electric timer and plug it into a wall outlet near your thermostat. Connect a small night-light to the timer and set the timer to turn the light on whenever you want the furnace off (and vice versa). Presto: When the little bulb is lit, it will produce just enough heat to fool the detector into thinking the entire living area is warm, and the furnace will switch off. Then when the preset timer deactivates the night-light, the thermostat will register the true temperature of the room and will turn the furnace on if the house is too cool.

Make sure the heat you pay for goes to warm you, not just your walls, ceilings, and furniture. If you have a radiator system, place a sheet of cardboard wrapped in foil behind each unit to reflect heat into the room. Floor registers can often be adjusted to direct hot air across the room rather than up (or you can fashion simple scooplike deflectors from triple thicknesses of foil and place one behind each grille).

You can put a stop to substantial heat loss by using shades and drapes, preferably lined or insulated, in the right places and at the right times. In winter, keep northern windows covered at all times, but open the shades and drapes on eastern and southern windows in the morning and close them late in the day. If it's a sunny afternoon, you can let sunlight in through western portals, too.

Here's a trick for those of you with double-hung windows that also have an exterior storm pane. In the fall, lower the upper sash far enough to pour plastic foam packing beads into the space between the panels. In the spring, just raise the lower sash and collect the material for use the next year. You won't want to do this with every window, because some—but not all—sunlight is blocked by the pellets. Start with a north window and use your own judgement as to how many subsequent units to fill.

Cracks under doors and around windowsills can help break your budget's back. One inexpensive way to stop these heat leaks is with tubes of cloth (discarded neckties are ideal) filled with sand. Sew one end of each cylinder shut so that you have a long "tube sock" about 4" in diam-

254

eter. Then pour in the granules, sew up the other end, and lay the draft dodger along the crevice.

Since moist air holds heat better than does dry air, you'll save money and stay warmer by keeping the relative humidity at 35% to 40%. A humidifier is the obvious (and costly) way to increase a home's moistness, but you can do almost as well with lots of healthy houseplants, or pans of water on radiators with—perhaps—a terry cloth towel hung above with the lower end touching the surface of the liquid.

To save a little extra of that power company "juice", get into the habit of turning lights off when you're not using them. Remember, too, that for brightly lit areas, one large incandescent bulb does a more efficient lighting job than will several small ones of the same total wattage. And when you're shopping for bulbs, compare the wattages and lumen values (actual light output) printed on the cartons. In many cases, you can get more light with less power simply by choosing one brand over another.

By using the appropriate clothing and bedding, you can be perfectly comfortable with a daytime room temperature setting of 65°F and one of 55°F at night. (Most authorities agree that cooler household temperatures are more healthful for the average fairly active individual.) Wear warm underclothing topped with one or more shirts and a sweater (the layered look is definitely in among conservationists). At night, several light blankets, rather than just one heavy coverlet, or a good down or fiberfill quilt will help keep you warm. So will long underwear, flannel sheets, furry critters, and a cuddlesome mate.

Use your appliances wisely and only when you must: Most dishwashers, dryers, freezers, and other household helpers perform much more efficiently when operated at full capacity. And apply these simple tips:

Oven/range: Keep your oven door closed while cooking, don't peek until food is done, but do let that hot air out into the room after the finished dish is removed. Remember, as well, that long preheat periods are seldom necessary when baking or broiling and that stovetop pots should be straight-sided, sized to fit their burners, and have tight-fitting lids.

Television: If you own an "instant on" type of TV (which is actually always operating), keep it unplugged until you want it on, or hook the unit into a socket that's controlled by a wall switch.

Refrigerator/Freezer: When you want something to eat, open the door, find and remove the item, and close the door quickly. In other words, don't dawdle (and teach your children not to, as well). In addition, keep the condenser coils at the back of your fridge clean. And defrost the freezer often: Just a quarter-inch of ice puts a heavy load on the appliance's motor.

Washer/Dryer: Wash clothes in lukewarm or cold water whenever possible. Keep lint filters, and the vent on your dryer, clean and unobstructed. And dry clothes in consecutive batches to take advantage of the heat buildup from previous loads. You could save a good deal of energy by hanging a small load of clothes on an indoor clothesline.

In any case, don't just throw up your hands in despair each winter and resign yourself to yet another season of outrageously high utility bills. There are steps you can take right now to cut your apartment's energy costs, but you're the one who must take action.

There are many easy ways a renter can reduce energy consumption, including "fooling" the thermostat with a timed night-light, making a radiator shield to direct warm air across the room, and insulating north-facing double-hung windows with plastic foam pellets.

AN ENERGY POTPOURRI

What's the Best Energy Saver for You?

The concern over the cost of heating a home—whether owned or rented—has led to the recognition that it's usually easier to use less energy than it is to produce, or pay for, more. Insulation, weatherization, and a plethora of other conservation techniques are valued and accepted ways to save on heating costs (and to help the worldwide energy bind) without sacrificing comfort. In general, most of these procedures could prove to be excellent investments, but on a personal basis, some are likely to be of greater financial benefit than others. There are several things you need to know before you can calculate just how much money can be saved by taking any given step toward conservation. The accompanying charts and the following explanation show you a method of determining which measures are best for you, your house, and your own particular climate.

The climate plays a direct role in determining whether or not adding insulation would be a wise investment. For convenience, the severity of a particular climate is usually stated as a number of heating *degree-days* (D-D). This figure is calculated by recording the average temperature for each day and then adding up the difference (in °F) between that figure and 65°F for each day on which the mercury readings average less than 65°. Thus (to take an example), if the temperature averages 50° for 100 days, the total number of degree-days for that period would be 1,500: 65–50 = 15 X 100 = 1,500 D-D. (If you adjust your thermostat higher or lower than 65°, your personal heating degree-day total will be proportionately larger or smaller.)

The accompanying map shows how the different degree-day regions spread out across the United States. You can estimate the degree-days in your hometown according to how far it's located from the nearest line. For example, St. Louis, Missouri lies between the lines indicating 4,000 and 6,000 degree-days, but it is somewhat closer to the 4,000 zone. Consequently, we can safely assume that city has between 4,500 and 5,000 heating degree-days (4,600 is the actual number).

Because the number of degree-days is proportional to the heat loss from your home, the figure is also directly related to the savings you can realize by installing additional insulation. To convert the Savings Factor in Chart 1 to your own potential saving in dollars per year, just multiply the Savings Factor by the first two digits of your degree-days, with a decimal point inserted so that the second digit is expressed as a fraction. With St. Louis's 4,500 degree-days, for example, use 4.5. Thus, the addition of 3-1/2″ of fiberglass insulation to an uninsulated attic in St. Louis would save 14¢ times 4.5, or 63¢ per square foot per year. If the house is 1,200 square feet, then the first year's saving would be $756.

Buying 1,200 square feet of 3-1/2″ fiberglass—at 18¢ per square foot—would require an expenditure of $216. Hence, the cost of putting that insulation in would be returned in just 3-1/2 months. The payback time can be calculated by dividing the Payback Factor (in this instance, 1.3) by the first digit and second-digit fraction of the degree-day total (again 4.5 in this example). The result will be expressed in tenths of a year.

Of course, the number of years that a conservation measure can be expected to last should play a role in your decision, too. Obviously, the longer the life expectancy of the insulation (or whatever), the more money it will save in the long

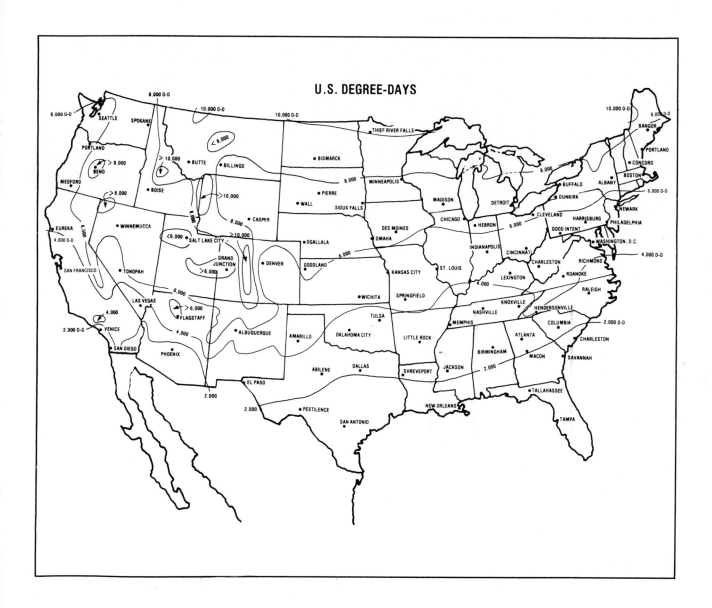

U.S. DEGREE-DAYS

AN ENERGY POTPOURRI

What's the Best Energy Saver for You? (continued)

CONSERVATION MEASURE	UNIT COST($)	SAVINGS FACTOR($)	PAYBACK FACTOR	LIFE EXPECTANCY
ATTIC INSULATION				
none (R-2.2) to 3-1/2" fiberglass (R-13.2)	0.18/ft.²	0.14	1.3	30
none (R-2.2) to 6" fiberglass (R-21.2)	0.30/ft.²	0.15	2.0	30
none (R-2.2) to 12" fiberglass (R-40)	0.60/ft.²	0.15	4.0	30
3-1/2" fiberglass (R-13.2) to 7" fiberglass (R-24.2)	0.18/ft.²	0.01	18.0	30
3-1/2" fiberglass (R-13.2) to 9-1/2" fiberglass (R-32.2)	0.30/ft.²	0.02	15.0	30
WALL INSULATION				
none (R-3.5) to 3-1/2" cellulose (R-16)	0.15/ft.²	0.08	1.9	20
none (R-3.5) to 5-1/2" cellulose[a] (R-22.8)	0.22/ft.²	0.09	2.4	20
3-1/2" fiberglass (R-13) plus 1" isocyanurate (R-22)	0.40/ft.²	0.01	40.0	30
3-1/2" fiberglass (R-13) plus 4" polystyrene (R-40) and stucco	2.00/ft.²	0.02	100.0	30
uninsulated block or brick wall (R-1.2) to 4" polystyrene (R-29.2) and stucco	2.00/ft.²	0.29	6.9	30
FLOOR INSULATION				
none (over closed, unheated space = R-8) to 3-1/2" fiberglass (R-19)	0.18/ft.²	0.03	6.0	30
none (over closed, unheated space = R-8) to 6" fiberglass (R-27)	0.30/ft.²	0.03	10.0	30
none (over open, unheated space = R-2.2) to 3-1/2" fiberglass (R-13.2)	0.18/ft.²	0.14	1.3	30
none (over open, unheated space = R-2.2) to 6" fiberglass (R-21.2)	0.30/ft.²	0.15	2.0	30

CONSERVATION MEASURE	UNIT COST($)	SAVINGS FACTOR($)	PAYBACK FACTOR	LIFE EXPECTANCY
BASEMENT INSULATION				
none (R-1.2) to 1" isocyanurate (R-20.2)	0.40/ft.²	0.03	13.3	30
CRAWL SPACE PERIMETER INSULATION				
none (R-1.2) to 3-1/2" fiberglass (R-12.2)	0.18/ft.²	0.27	0.7	30
none (R-1.2) to 6" fiberglass (R-20.2)	0.30/ft.²	0.28	1.1	30
STORM WINDOWS				
1-glaze (R-0.9) to 2-glaze (R-1.9)	4.00/ft.²	0.21	19.0	30
1-glaze (R-0.9) to 3-glaze (R-2.9)	8.00/ft.²	0.28	28.6	30
2-glaze (R-1.9) to 3-glaze (R-2.9)	4.00/ft.²	0.07	57.1	30
INSULATING SHUTTERS				
1-glaze (R-0.9) to R-6	4.00/ft.²	0.34	11.8	30
2-glaze (R-1.9) to R-6	4.00/ft.²	0.13	30.8	30
ELECTRIC WATER HEATER (@ 5¢/KWH) INSIDE				
setback from 150°F to 120°F	0.00	8.71[c]		
add R-4 blanket to make R-8 heater @ 150°F	20.00[b]	12.34[c]	1.6	30
add R-4 blanket to make R-8 heater @ 120°F	20.00[b]	7.98[c]	2.5	30
FURNACE AIR DUCT INSULATION (IN UNHEATED SPACE)[d]				
none (R-1) to 2" fiberglass (R-7)	0.22/ft.²	0.85/ft.²/ 100 gal. fuel oil	0.3	20
WEATHERSTRIP AND CAULK				
old house (3 air-changes/hr.) to 1/2 air-change/hr.	0.10/ft.² of floor space	0.15/ft.² of floor space	0.3	10
new house (1.5 air-changes/hr.) to 1/2 air-change/hr.	0.07/ft.² of floor space	0.06/ft.² of floor space	1.2	10

[a] original walls with 6" studs required to accommodate 5-1/2" blown cellulose   [b] total cost
[c] same in all zones   [d] assumptions: 120°F forced air, 30°F ambient air (miscellaneous variables make figures approximate)

CHART 1

run. However, many folks will move to a new home before any of the materials listed lose their effectiveness and thus may not get a return on their investment for the full life of the installation. On the other hand, a well-insulated house is almost certain to have a better resale value than does a comparable dwelling that hasn't been properly weatherized.

You now have the basic tools to choose the appropriate energy-saving methods for you. But before making final decisions, it's important to understand the basis of this analysis ciphering and to be certain the method fits your own situation.

Be sure that the method of insulating you're considering will be a practical one in your home. Adding thick exterior insulation to a frame house with siding, for example, isn't an easy procedure: The siding might have to be removed and sheathing put up. Or, if your dwelling doesn't have an attic, the standard fiberglass blanket insulation listed won't be applicable: You'd need to use more expensive rigid insulation or lower the interior ceilings. There are hundreds of different ways to insulate, and you need to be well aware of the effort, expense, and savings that go with those you're considering.

The dollar values used in this analysis were reached by averaging 1982 costs for materials and energy, and the selection of materials and determination of the insulative values (R-values) of the different insulation techniques were based on standard construction techniques. The costs will undoubtedly vary throughout the country and fluctuate with time. Also, in all but one case, the cost figures here are based only on the price of the materials needed to do the job. The work involved in the methods listed should all be within the abilities of a handyperson in possession of a good how-to book (except for the stucco portion of the 4"-polystyrene exterior insulation . . . and in that case, $1.00 per square foot was added to cover the cost of professional stucco or surface-bonding cement work). It's your labor that will make possible the rapid economic returns described here, and if you figure an hourly rate on the basis of the lifetime saving achieved, you should be quite pleased with your pay scale.

The savings and payback factors claimed are based upon the use of fuel-oil heating at $1.25 per gallon, burned in a furnace that's 65% efficient. If you depend on another method—or have one of the new, more efficient flame-retention-head burners on your oil furnace—the savings and payback factors may prove to be greater. Chart 2 lists the common means of heating a house and the approximate cost per 1,000 British Thermal Units (Btu) delivered. You can use the numbers to recalculate your own savings and payback factors . . . provided you're paying about the same price as specified for the particular heat source.

For example, if you heat with electricity, the cost per 1,000 Btu is just about the same as that of oil, if you pay 5¢ per kilowatt-hour (kwh). At 10¢ per kwh, electricity costs about twice as much as fuel oil does per 1,000 Btu. In such a case, the savings would be doubled, and the conservation measure would pay for itself in half the time.

The final possible discrepancy to watch out for in the figures is the beginning R-value listed in Chart 1. Differences in wall, ceiling, and floor construction techniques can alter uninsulated R-values by 1 or more. If your building has a lower

CHART 2				
HEAT SOURCE	BTU/UNIT	COST/UNIT($)	EFFICIENCY(%)	COST/1,000 BTU($)
oil	131,000/gal.	1.25/gal.	65	0.015
electricity	3,410/kwh	0.05/kwh	100	0.015
natural gas	1,000/ft.³	0.005/1,000 ft.³	70	0.007
coal	25 X 10⁶/ton	125/ton	60	0.008
wood	28 X 10⁶/cord	100/cord	50	0.007
solar	3 X 10⁶/ft.²/ 20-yr. lifetime	15/ft.²	100	0.004–0.008

AN ENERGY POTPOURRI

What's the Best
Energy Saver for You? (continued)

or higher original R-value than shown, the savings and payback factors will be affected. Consequently, you should either consider the numbers to be approximations or do your own careful calculations, following the method given for calculating savings and payback.

With these easy methods at your disposal, you should be able to make sound decisions about what conservation techniques make the most sense for you. Of course, you simply may not have enough cash available to take the step with the best payback, so you might have to opt for one that requires a lower initial investment. Furthermore, you may feel more comfortable with your ability to perform some jobs than others. Remember, the ratings provided are merely suggestions designed to help the homeowner make a decision.

No standard investment—whether it be a money-market certificate, bond, stock, or collectible—is likely to pay you so much for such a small outlay. Plus, even without the tax credits available in many states and from the federal government, the money you save by conserving is the equivalent of tax-free income. Can you afford *not* to save energy?

CALCULATING SAVINGS AND PAYBACK

The savings and payback factors shown in Chart 1 can be used to get a good idea of the relative savings and payback periods for numerous weatherizing and insulating methods. People who have narrowed the field of alternatives to be considered will probably want to make more specific calculations geared to their specific dwelling. This can be easily done by using the following formula.

STEP 1: $\dfrac{1}{\text{old R-value}}$ = loss in Btu/ft.²/hr. (old U-value)

STEP 2: result of Step 1 X degree-days X 24 hours = old loss in Btu/ft.²/yr.

STEP 3: $\dfrac{1}{\text{new R-value}}$ = new loss in Btu/ft.²/hr. (new U-value)

STEP 4: result of Step 3 X degree-days X 24 hours = new loss in Btu/ft.²/yr.

STEP 5: result of Step 2 − result of Step 4 = Btu/ft.²/yr. saved

STEP 6: result of Step 5 X $/Btu = $/ft.²/yr. saved

STEP 7: $\dfrac{\$/\text{ft.}^2\text{ cost}}{\$/\text{ft.}^2/\text{yr. saved}}$ = PAYBACK (in years)

STEP 8: $/ft.²/yr. saved X total ft.² = SAVINGS/yr.